The Palgrave Macmillan Animal Ethics Series

Series Editors

Andrew Linzey
Oxford Centre for Animal Ethics
Oxford, UK

Clair Linzey
Oxford Centre for Animal Ethics
Oxford, UK

In recent years, there has been a growing interest in the ethics of our treatment of animals. Philosophers have led the way, and now a range of other scholars have followed from historians to social scientists. From being a marginal issue, animals have become an emerging issue in ethics and in multidisciplinary inquiry. This series will explore the challenges that Animal Ethics poses, both conceptually and practically, to traditional understandings of human-animal relations. Specifically, the Series will:

- provide a range of key introductory and advanced texts that map out ethical positions on animals;
- publish pioneering work written by new, as well as accomplished, scholars;
- produce texts from a variety of disciplines that are multidisciplinary in character or have multidisciplinary relevance.

For further information or to submit a proposal for consideration, please contact Amy Invernizzi, amy.invernizzi@palgrave-usa.com.

Aude-Solveig Epstein • Alice Di Concetto
Editors

Animals in EU Economic Law

A New Approach to Protecting Sentient Beings

palgrave
macmillan

Editors
Aude-Solveig Epstein
Sorbonne University Abu Dhabi
Abu Dhabi, United Arab Emirates

Alice Di Concetto
The European Institute for Animal Law & Policy
Brussels, Belgium

ISSN 2634-6672 ISSN 2634-6680 (electronic)
The Palgrave Macmillan Animal Ethics Series
ISBN 978-3-032-16170-3 ISBN 978-3-032-16171-0 (eBook)
https://doi.org/10.1007/978-3-032-16171-0

This Palgrave Macmillan imprint is published by the registered company Springer Nature Switzerland AG.
The registered company address is: Gewerbestrasse 11, 6330 Cham, Switzerland

"This book brilliantly shows how the harm done to animals is not a mere 'externality' to the working of markets, but forms part of the legal regimes on trade, competition, consumer protection, procurement, and finances. Making concern for animal flourishing an objective of those legal regimes is a key contribution to the paradigm shift towards a sustainable economic law."

–Anne Peters, Director at the *Max Planck Institute for Comparative Public Law and International Law Heidelberg*, Germany

"*Animals in Economic Law* looks beyond standard animal law paradigms to offer a pathbreaking exploration of how markets could be reconceived and restructured to protect the interests and dignity of animals. This novel and timely contribution should be read by all scholars and practitioners concerned with the well-being of nonhuman animals."

—Douglas Kysar, Joseph M. Field '55 Professor of Law, Faculty Director, Law, Environment, and Animals Program, *Yale Law School*

"Could the key to animal welfare lie not in animal law, but in economic law? This conundrum is brilliantly explored by this book which reimagines the foundations of animal welfare. A landmark contribution to animal law studies".

—Nitish Monebhurrun, Law Professor (*University Center of Brasília*, Brazil), Arbitrator (CAMESC/ALAM), Expert (*Division for Ocean Affairs and the Law of the Sea*, United Nations)

"This book offers timely and penetrating arguments on the reforms necessary to advance the holistic implementation of animal law and policy, including the obligation of business enterprises and other economic actors to promote animal rights across their value chains. Alice Di Concetto and Aude-Solveig Epstein must be commended for shedding light on this important, but still largely understudied area of law."

—Damilola S. Olawuyi SAN, Independent Expert, United Nations Working Group on Business and Human Rights, Professor and UNESCO Chair on Environmental Law and Sustainable Development, *Hamad Bin Khalifa University*

"This outstanding edited volume shows that meaningful progress for animals is possible within the existing framework of EU economic law. The authors address a broad range of issues from both production and consumption perspectives, ultimately setting a new research agenda for protecting animals as sentient beings across Europe and beyond."

—Vincent N. Delhomme, *Assistant Professor of Law, Leiden University*, The Netherlands

Series Editors' Preface

This is a new book series for a new field of inquiry: Animal Ethics.

In recent years, there has been a growing interest in the ethics of our treatment of animals. Philosophers have led the way, and now a range of other scholars have followed, from historians to social scientists. From being a marginal issue, animals have become an emerging issue in ethics and in multidisciplinary inquiry.

In addition, a rethink of the status of animals has been fuelled by a range of scientific investigations which have revealed the complexity of animal sentiency, cognition and awareness. The ethical implications of this new knowledge have yet to be properly evaluated, but it is becoming clear that the old view that animals are mere things, tools, machines or commodities cannot be sustained ethically.

But it is not only philosophy and science that are putting animals on the agenda. Increasingly, in Europe and the United States, animals are becoming a political issue as political parties vie for the "green" and "animal" vote. In turn, political scientists are beginning to look again at the history of political thought in relation to animals, and historians are beginning to revisit the political history of animal protection.

As animals grow as an issue of importance, so there have been more collaborative academic ventures leading to conference volumes, special journal issues, indeed new academic animal journals as well. Moreover, we have witnessed the growth of academic courses, as well as university posts, in Animal Ethics, Animal Welfare, Animal Rights, Animal Law, Animals and Philosophy, Human-Animal Studies, Critical Animal Studies, Animals and Society, Animals in Literature, Animals and Religion—tangible signs that a new academic discipline is emerging.

"Animal Ethics" is the new term for the academic exploration of the moral status of the non-human—an exploration that explicitly involves a focus on what we owe animals morally, and which also helps us to understand the influences—social, legal, cultural, religious and political—that legitimate animal abuse. This series explores the challenges that Animal Ethics poses, both conceptually and practically, to traditional understandings of human–animal relations.

The series is needed for three reasons: (1) to provide the texts that will service the new university courses on animals; (2) to support the increasing number of students studying and academics researching in animal related fields; and (3) because there is currently no book series that is a focus for multidisciplinary research in the field.

Specifically, the series will

- provide a range of key introductory and advanced texts that map out ethical positions on animals;
- publish pioneering work written by new, as well as accomplished, scholars, and
- produce texts from a variety of disciplines that are multidisciplinary in character or have multidisciplinary relevance.

The new Palgrave Macmillan Series on Animal Ethics is the result of a unique partnership between Palgrave Macmillan and the Ferrater Mora Oxford Centre for Animal Ethics. The series is an integral part of the mission of the Centre to put animals on the intellectual agenda by facilitating academic research and publication. The series is also a natural complement to one of the Centre's other major projects, the *Journal of Animal Ethics.* The Centre is an independent "think tank" for the advancement of progressive thought about animals, and is the first Centre of its kind in the world. It aims to demonstrate rigorous intellectual enquiry and the highest standards of scholarship. It strives to be a world-class centre of academic excellence in its field.

We invite academics to visit the Centre's website www.oxfordanimalethics.com and to contact us with new book proposals for the series.

Oxford, UK

Andrew Linzey
Clair Linzey

Acknowledgments

This book is the result of several years of reflection on the place of animals in the legal system, particularly within economic law. The editors and contributors first presented these reflections at a conference held on March 24–25, 2023, in Paris, hosted by the *Institut d'Études Avancées* de Paris and Sciences Po Law School. The conversations begun there have matured over time and culminated in the publication of this book. We are deeply grateful to all those whose efforts helped bring this project to fruition.

We would like to extend our heartfelt thanks to Reagan M. Sova for his invaluable editorial work. His guidance and expertise helped us find the right words with precision and nuance. We are also grateful to Gabriela Kubíková, Gladys Le Goff, and Anatole Poinsot for their outstanding research and editing skills, as well as to Pauline Koczorowski for her contributions on the topic of EU and French aquatic animal law. We are equally thankful to Qinci Li for assisting us with her rigorous and meticulous formatting work.

We additionally extend our gratitude to all those who helped us organize the conference in 2023, especially Sandrine Morvan at the IEA de Paris. We acknowledge the generosity of Sciences Po Law School, which welcomed us on many occasions and provided space for these ideas to take root. We also thank the different programs and research groups who provided essential support for this project: the *Centre d'études juridiques européennes et comparées* (University of Paris Nanterre), the *Centre de droit civil des affaires et du contentieux économique* (University of Paris Nanterre), the *Laboratoire Droit & Changement Social* (University of Nantes), and the *Groupe de Recherche en Droit, Économie et Gestion* (University of Nice).

We are grateful to Andrew and Clair Linzey for supporting this project from its inception and for helping us find the right home for the publication of this book. A sincere thank you as well to the team at Palgrave Macmillan team, especially Amy Invernizzi and Antony Sami, for their patience and professionalism.

We are also indebted to the TEDE research group for their foundational reflections on the ecological turn in economic law. Their incisive analysis of how economic law can advance sustainability laid a robust foundation for this volume. Lastly, we also thank the scholars who generously reviewed the manuscript: Doug Kysar, Nitish Mohneburrun, Damilola Olawuyi, and Anne Peters.

Contents

Notes on Contributors

Iyana Aït Ahmed graduated from Sciences Po Paris Law School and is a lawyer at the Paris Bar. Her research interests include human rights and animal law. Together with Solène Kerisit, she led a project at the Science Po Law School Legal clinic, under the supervision of Professor Aude-Solveig Epstein and Alice Di Concetto. The project examined the legal framework applicable to farmers in France, assessing its role in farmers' precariousness, loss of autonomy, and work-related suffering—as well as on the treatment of farmed animals.

Clémentine Baldon is a lawyer at the Paris Bar. She founded Baldon Avocats, a law firm which operates at the intersection of trade, sustainability, and regulatory compliance, focusing on high-impact litigation and advisory work in EU law, environmental law, and international trade. Her key areas of work include strategic litigation and regulatory challenges before French and EU courts; international trade, WTO, and EU law compliance; investment arbitration; corporate accountability and sustainable public procurement.

Sophie Bresny was appointed General Rapporteur of the Polynesian Competition Authority in May 2023, after leading the investigation department of the French National Competition Authority in mainland France. Before joining the French National Competition Authority, she worked at the *Direction générale de la concurrence, de la consommation et de la répression des fraudes* (DGCCRF), the French agency tasked with implementing competition and consumer protection law, where she led telecommunications cases for nearly three years. Between 2009 and 2010,

Bresny worked in Dakar, Senegal, for the US nonprofit Tostan, which promotes education programs in communities across West Africa.

Nicolas de Sadeleer is Full Professor at UCLouvain, Saint-Louis. He specializes in EU law (institutions, internal market), environmental law (international and domestic), and comparative law. He is an active commentator on EU legal and political issues in the areas of trade, investment, and sustainable development. In addition to holding guest academic positions at over forty universities around the world, he has been the recipient of six international university chairs. In addition, de Sadeleer has worked as a lawyer and as a consultant with national and international authorities on a wide range of environmental issues. He is supervising several research projects on the implementation of the EU Green Deal. His research has been published by leading scholarly publishing houses and journals globally.

Alice Di Concetto is the Executive Director and Chief Legal Advisor of the European Institute for Animal Law & Policy (Brussels, Belgium). She also lectures on EU animal ethics at Sciences Po (Paris, France). From 2016 to 2018, she served as a fellow in the Animal Law & Policy Program at Harvard Law School. Her publications include several law review articles, and she is the author of the forthcoming book EU Animal Law (Edward Elgar Publishing, 2026). Di Concetto earned a Master's degree in Animal Law (LL.M, 2016) from Lewis & Clark Law School (USA), for which she obtained a Fulbright grant. She graduated from Sciences Po Law School in Paris (Master in Law, 2015).

Isabelle Doussan is a research director at INRAE in the *Groupe de Recherche en Droit, Economie, Gestion* in Nice, France. A photographer turned lawyer, Doussan started her academic career as an Assistant professor (*Maîtresse de conférences*) at the University of Rennes Faculty of Law. Her research initially focused on agriculture and environmental law, and later expanded to include the legal analysis of complex notions such as ecosystem services and functions, ecological compensation, and human–non-human relationships. Doussan's work examines how the law addresses the relationship between agricultural activities and the environment, and more broadly, non-human life.

Aude-Solveig Epstein is the Provost of Sorbonne University Abu Dhabi. She previously served as Visiting Associate Professor of Legal Studies at NYU Abu Dhabi and Associate Professor of Law at Paris Nanterre

University. Her research sits at the intersection of economic, environmental, and animal law, exploring how legal frameworks can guide economic systems toward sustainability. Since 2021, she has co-led the TEDE research group on the ecological transition of economic law. She co-edited *Environmental Economic Law: Actors and Methods* (Mare & Martin, 2023) and *Can Economic Law Steer the Ecological Transition?* (Bruylant, 2022). She is also the lead coordinating author of *The Ecological Transformation of Economic Law* (Mare & Martin, forthcoming, 2026). Her monograph on *Corporate Environmental Disclosure: A Legal Analysis* (L.G.D.J., 2015) received the Louis Joinet Prize.

Marine Friant-Perrot is *Maître de conférences* at the School of Law and Political Sciences at the University of Nantes. She is qualified to supervise doctoral and postdoctoral researchers. She has published extensively in peer-reviewed journals in French and English on a broad range of consumer protection and food law issues, as well as on how these two disciplines relate to public health.

Sophie Grosbon is a Professor of international law and a researcher at *CEDIN*, Paris Nanterre University. Her work explores the interactions between international trade law, international environmental law, and the protection of human rights. Her contribution to this volume was made as part of a *Centre national de la recherche scientifique* delegation to DICE-CERIC at Aix-Marseille University.

Solène Kerisit is studying at the Québec Bar School. She holds a degree in Economic Public Law from Sciences Po Law School and an LLM in public law from the University of Montréal Faculty of Law. Kerisit has worked in French and Canadian law firms specialized in environmental and public law and has participated in research projects dedicated to intensive farming, environmental law, and fundamental rights in France, the United States, and Canada.

Anne-Laure Meynckens is a lawyer specializing in public law and the founder of the consultancy *Drôle De Zèbre*. She worked for ten years at the French Ministry of the Economy as an inspector for the *Direction générale de la concurrence, de la consommation et de la répression des fraudes* (DGCCRF), the agency tasked with implementing competition and consumer protection law. With a background in human–animal relations, Meynckens founded *Drôle De Zèbre* in 2020, providing training and consultancy services on animal-related issues and assisting local authorities in developing animal welfare policies. She is also the vice president of the nonprofit *Éducation Éthique Animale* and the co-founder of the Agence Symbiocène.

Denys-Sacha Robin is an Assistant Professor at Paris Nanterre University, a member of the Center of International Law, and Director of a Specialized Diploma in Law of Energies, International Investment, and Arbitration. His main publications concern general international law (sources, state responsibility), international law of the sea (maritime disputed zones, maritime biodiversity), and international animal law. He is currently focusing on the international law of natural resources and energy transition.

Introduction

Alice Di Concetto and Aude-Solveig Epstein

Is animal liberation achievable within the framework of financial capitalism? The persistent failure of laws and policies in capitalist systems to eliminate even the most extreme forms of animal cruelty raises serious doubts. A prime example is so-called farm animal welfare legislation, which typically accompanies the ongoing intensification of factory farming, further entrenching the industrial violence visited upon animals, as opposed to effectively limiting it.[1] More generally, the effectiveness of animal welfare laws in protecting animals is so minimal that animal welfare legislation is

[1] Francione, G. 1995. *Animals, Property, and the Law*. Philadelphia: Temple University Press.

A. Di Concetto (✉)
The European Institute for Animal Law & Policy, Brussels, Belgium
e-mail: contact@animallaweurope.org

A.-S. Epstein
Provost at Sorbonne University Abu Dhabi (CEDCACE/SAFIR),
Abu Dhabi, UAE
e-mail: aude.epstein@sorbonne.ae

A.-S. Epstein, A. Di Concetto (eds.), *Animals in EU Economic Law*, The Palgrave Macmillan Animal Ethics Series,
https://doi.org/10.1007/978-3-032-16171-0_1

often indistinguishable from the legislative framework that enables animal exploitation.[2]

Few jurisdictions around the world have enacted legislation that effectively shields animals from mass-scale exploitation, and virtually no major economic power has passed adequate animal protection laws. The European Union (EU), which often portrays itself as a leader on animal protection issues, is no exception. Since the late 1970s, the EU has enacted animal protection legislation in a variety of fields such as farmed animal welfare,[3] the regulation of wildlife trade,[4] and the treatment of animals used for scientific purposes.[5] Yet these laws, which borrow from the vocabulary of traditional animal welfarism, have done little to curb cruel business practices, which continue unabated. Even the EU's trade bans on cruel wild animal products, adopted in 1991[6] and 2009,[7] stand as symbolic gestures rather than effective interventions, having no discernible impact on industrial practices at play within the EU.[8]

The EU recently announced major reforms that appeared to depart from the pattern of minimalist change that has long characterized its animal law and policy. In 2020, the European Commission—the EU's executive branch—announced a reform of EU farm animal welfare "to align it with the latest scientific evidence, broaden its scope, make it easier to enforce and ultimately ensure a higher level of animal welfare."[9] In 2021, the Commission further announced a ban on the use of cages in animal

[2] Stucki, S. 2023. Animal Warfare Law and the Need for an Animal Law of Peace: A Comparative Reconstruction. *The American Journal of Comparative Law* 71(1): 189–233.

[3] For instance, Council Directive 98/58/EC of 20 July 1998 concerning the protection of animals kept for farming purposes, 1998 O.J. (L 221) 23–27.

[4] For instance, Council Regulation (EC) No 338/97 of 9 December 1996 on the protection of species of wild fauna and flora by regulating trade therein, 1996 O.J. (L 061).

[5] Directive 2010/63/EU of the European Parliament and of the Council of 22 September 2010 on the protection of animals used for scientific purposes, 2010 O.J. (L 276) 33–79.

[6] Council Regulation (EEC) No 3254/91 of 4 November 1991 prohibiting the use of leghold traps in the Community and the introduction into the Community of pelts and manufactured goods of certain wild animal species originating in countries which catch them by means of leghold traps or trapping methods which do not meet international humane trapping standards, 1991 O.J. (L 308) 1–4.

[7] Regulation (EC) No 1007/2009 of the European Parliament and of the Council of 16 September 2009 on trade in seal products, 2009 O.J. (L 286) 36–39.

[8] European Commission. 2022. Commission Staff Working Document Fitness Check of the EU Animal Welfare legislation. SWD (2022) 328 final.

[9] European Commission. 2020. Communication from the Commission to the European Parliament, the Council, the European Economic and Social Committee and the Committee

agriculture, following a European Citizens' Initiative signed by nearly 1.4 million citizens calling for an "end [to the] cage age."[10] Yet five years later, EU farm animal welfare legislation remains virtually unchanged. The Commission has quietly walked back its commitments, and no tangible progress is in sight. These failed reform efforts suggest that animal exploitation is ingrained in our economies to a greater extent than reliance on fossil fuels, causing optimism to be "less justified for the protein transition than for the energy transition."[11] More broadly, this situation reinforces the view that reformism has failed animals.[12]

At the same time, proponents of animal rights law theory, who argue that no significant change in human–animal relations can be achieved unless animals are afforded legal personhood and fundamental rights,[13] have equally failed. Animal rights theory has not found its way into legislative advocacy, on account of the chasm between the solutions of animal rights advocates and citizens' aspirations.[14] Animal rights advocates have equally failed in courts, which generally refuse to grant personhood to animals,[15] and when they do, the resulting rulings often offer practical

of the Regions, A Farm to Fork Strategy for a fair, healthy and environmentally-friendly food system. COM(2020) 381 final, p. 8.

[10] European Commission. 2021. European Citizens' Initiative: Commission to propose phasing out of cages for farm animals. Available at: https://ec.europa.eu/commission/presscorner/detail/en/ip_21_3297

[11] Noelkaemper, A. 2025. The Other 'Transitioning Away' Imperative: Meat as the Next Frontier in Global Climate Change Policy. EJIL Talk! 7 January. Available at: https://www.ejiltalk.org/the-other-transitioning-away-imperative-meat-as-the-next-frontier-in-global-climate-change-policy/

[12] Francione, G. L. 1996. Animal Rights and Animal Welfare. *Rutgers Law Review* 48(2): 97–470.

[13] Steven, M. W. 2000. *Rattling the Cage: Toward Legal Rights for Animals.* Cambridge, Massachusetts: Perseus Books.

[14] Lovvorn, J. R. 2006. Animal Law in Action: The Law, Public Perception, and the Limits of Animal Rights Theory as a Basis for Legal Reform. *Animal Law Review* 133.

[15] To date, only one court, in Argentina, recognized that an animal was a person and was therefore entitled to the fundamental right of habeas corpus. Tercer Juzgado de Garantías de Mendoza [J.G.Men.] [Third Criminal Court of Mendoza], 3 November 2016, Presentación Efectuada Por AFADA Respecto del Chimpancé "Cecilia" Sujeto No Humano, [Expte. Nro.] P-72.254/15 (Argentina). For a review of cases seeking the recognition of animal personhood, see Franceschini, M. M. 2021. Animal Personhood: The Quest for Recognition. *Animal & Natural Resource Law Review.*

solutions that could have been achieved without invoking legal personhood.[16]

Animal protection efforts are thus caught in a conundrum, prompting a growing number of scholars to explore the potential of alternative legal categories, beyond the property/person binary,[17] that could support a progressive shift from animal welfare to animal liberation.[18]

The contributions to this book suggest that there is another, potentially complementary, legal approach to overcoming these theoretical and practical deadlocks. Taken together, these contributions hypothesize that significant shortcomings in EU animal protection law stem from a disconnect between animal protection statutes and economic law, in a way that prioritizes narrowly defined economic objectives over animal protection goals. In the context of this book, the term "economic law" designates the rules governing companies and markets, including but not limited to corporate law, financial regulation, consumer law, competition law, public procurement law, and international trade law.

The approach presented in this book thus advocates for reshaping the foundations of economic law, so as to bring coherence between economic interests and animal protection objectives. This approach is consistent with Sue Donaldson and Will Kymlicka's view that the pursuit of economic growth and animal exploitation are not "an inevitable unfolding of a logic of capitalism or enlightenment rationalism. These developments were the (frustrating, infuriating) outcome of political contestations, not the 'logic' of liberalism."[19] To remedy these developments, the contribu-

[16] The remedies in the Argentinian rulings are illustrative from that perspective. Remedies, such as the transfer of an orangutan to a sanctuary, do not require the recognition of animal personhood. The demonstration that the animal is exposed to abuse can be enough for the courts to remove an animal from abusive situations, as is regularly ruled in jurisdictions that do not recognize animal personhood. More broadly, the same critique of the inability of Rights of Nature to protect the environment can be formulated against animal rights theory and its inability to effectively protect animals. For a critique of Rights of Nature, see Bétaille, J. 2019. Rights of Nature: Why it Might Not Save the Entire World? *Journal for European Environmental & Planning Law* 16, p. 35.

[17] Kymlicka, W. 2017. Social membership: animal law beyond the property/personhood impasse. *Dalhousie Law Journal* 40(1): 123–156.

[18] Kempers, E. B. 2022. Transition Rather than Revolution: The Gradual Road towards Animal Legal Personhood through the Legislature. *Transnational Environmental Law* 581.

[19] Donaldson, S., and W. Kymlicka. 2016. Make It So: Envisioning a Zoopolitical Revolution. In *Philosophy and the Politics of Animal Liberation*, ed. Paola Cavalieri, 71–116. New York: Palgrave Macmillan.

tors to this book argue that integrating animal protection objectives into economic law could advance the interests of animals, enhance legal coherence, and promote economic efficiency.

Animal protection issues have attracted increasing scholarly interest in recent years, in a variety of fields, including competition law,[20] commercial law,[21] international trade law,[22] consumer law,[23] and financial law.[24] However, this scholarly work has developed in a fragmented manner, lacking comprehensive transdisciplinary anchoring and a coherent theoretical framework. Furthermore, contributions rooted in a European context remain sparse. This book attempts to bridge these gaps.

The book opens with the first contribution, Animals in Economic Law: A Research Agenda, by editors Aude-Solveig Epstein and Alice Di Concetto, who situate the volume at the intersection of economic law and animal law. They argue that the dominant legislative approach – developing animal law in isolation from economic law – has limited its transformative potential and its ability to tackle the root causes of systemic animal exploitation. In response, the authors advocate for a more integrated approach, embedding animal protection objectives within economic law. This integration, they contend, would reshape how economic value is assessed in a way that better aligns the interests of economic decision-makers with those of the animals affected by their choices.

The first section of the book, entitled "Animals as Economic Input," brings together four contributions that examine how different areas of economic law treat animals as productive inputs, and explore potential avenues for reform within these supply-side legal frameworks. In their contribution "The Common Agricultural Policy and Animals: Friends or Foes?" Nicolas de Sadeleer and Alice Di Concetto analyze current EU agricultural subsidy rules and present reform proposals to compel the

[20] See, e.g., Sluszka, D. 2016. Animal Farm: The E.U.'s Move towards Progress and the U.S.'s Slide towards Dystopia in Farm Animal Welfare. *Journal of International and Comparative Law* 24: 423–454.

[21] See, e.g., Thomas, N., ed. 2022. *Animals and Business Ethics*. New York: Palgrave Macmillan.

[22] See, e.g., Blattner, C. E. 2019. Beyond the Goods/Resources Dichotomy: Animal Labor and Trade Law. *Journal of International Wildlife Law & Policy* 22(2): 63–89.

[23] See, e.g., Peters, A. 2021. *Animals in international law*. Leiden: Brill-Nijhoff, 108 sq.; Di Concetto, A., and A.-S. Epstein. 2023. EU Consumer Information as a Tool to Regulate the Treatment of Farm Animals. *EU Journal of Consumer Law*: 145–174.

[24] Peters, A. 2021. *Animals in international law*. Leiden: Brill-Nijhoff, 97 sq.

transition to more humane and sustainable farming systems. Expanding the perspective on the EU's food system, Denys-Sacha Robin's contribution, "Protecting Marine Life: Legal Gaps and Policy Contradictions in EU Fishing Regulations," offers a critical analysis of the shortcomings in EU fisheries regulations when it comes to protecting marine ecosystems. Robin argues that these deficiencies result from EU institutions prioritizing the short-term financial interests of large-scale fisheries operators.

Focusing on the situation of farmers, Ilyana Aït-Ahmed and Solène Kerisit's contribution, "Cruelty Under Contract: A Case Study in French Law," traces the shared roots of animal exploitation and farmers' disempowerment to the legal norms that structure agri-food value chains. Lastly, Clémentine Baldon's contribution, "Corporate Social Responsibility and Animal Protection," examines current trends, challenges, and achievements related to the integration of animal protection objectives into corporate social responsibility standards. She also explores the potential for advancing these objectives through strategic litigation.

The second and final section of the book, entitled "Animals as Consumption Goods," examines the condition of animals through the lens of economic law areas that treat them as consumption goods, and explores potential reform avenues from within these demand-side policies. This section starts with a contribution by Sophie Bresny, "What Can Competition Law Do for Animals? Recent Developments and Future Directions Under the French Competition Authority," on the synergies between antitrust law and animal protection. "Animal Protection in Public Procurement Rules," by Anne-Laure Meynckens, explores the inclusion of animal protection standards in public procurement rules as a potentially powerful measure that is currently under-utilized. This contribution further presents the potential challenges of enhancing such standards, and the positive effects such enhanced standards might have on the treatment of animals. Turning to the protection of wild animals, Sophie Grosbon, in "The Protection of Endangered Species in EU and International Trade Law," shows that while international laws aimed at protecting animals through trade regulation have limited effects, the implementation of such laws at regional and national levels can pave the way to much more ambitious levels of animal protection.

The two contributions that conclude the book explore consumer protection laws as a potential lever to advance the protection of animals. Isabelle Doussan, in "Using Consumer Law to Improve the Treatment of Animals in Industrial Farming Production," offers a critical perspective on

EU animal welfare standards and suggests solutions to alleviate animal suffering, including through better consumer protection rules. Similarly, Marine Friant-Perrot, in "Meat Reduction as an Animal Protection Policy Goal: Exploring the Economic, Health, and Climate Nexus in EU Law," analyzes the many ways in which the EU legislature could reform the law to initiate the transition to a more plant-based economy.

Animals in Economic Law: A Research Agenda

Aude-Solveig Epstein and Alice Di Concetto

1 Introduction

In a context where animal law and animal ethics often diverge, a compelling new approach to animal advocacy is gaining traction. It involves leveraging market regulations and business law (economic law) to promote animal interests, either as an alternative to or in conjunction with traditional animal protection statutes.

Over the years, animal advocacy organizations have increasingly turned to corporate-based campaigns to encourage managers to adopt better animal welfare practices. These efforts have included the development of

A.-S. Epstein (✉)
Provost at Sorbonne University Abu Dhabi (CEDCACE/SAFIR), Abu Dhabi, UAE
e-mail: aude.epstein@sorbonne.ae

A. Di Concetto
The European Institute for Animal Law & Policy, Brussels, Belgium
e-mail: contact@animallaweurope.org

A.-S. Epstein, A. Di Concetto (eds.), *Animals in EU Economic Law*, The Palgrave Macmillan Animal Ethics Series,
https://doi.org/10.1007/978-3-032-16171-0_2

food labels,[1] cage-free pledges,[2] and broader animal welfare commitments.[3] In parallel, animal advocates have engaged in shareholder activism, drawing inspiration from strategies previously employed by environmental advocates.[4] Additionally, in the context of factory farming, there has been a notable shift away from pursuing individual convictions for animal cruelty–an approach often criticized for its limited systemic impact and for disproportionately affecting marginalized communities.[5] Instead, advocacy groups now insist on holding corporations, whose business models rely primarily on animal exploitation, along with their executives, to account for animal cruelty across their value chains. To that end, advocates have notably challenged the legality of misleading advertising, targeting both consumers[6] and investors.[7]

This shift reflects a creative strategy by animal advocates to achieve incremental change for animals while circumventing the traditional roadblocks of legislative reform.[8] It also underscores how market protection goals are prioritized over animal protection within legal systems. Because legislatures tend to view harm to the market as more serious than cruelty

[1] For more on this, see Di Concetto, A. and A.-S. Epstein. 2023. EU Consumer Information as a Tool to Regulate the Treatment of Farm Animals. *EU Journal of Consumer Law*: 145–174; Di Concetto, A. 2023. Farm Animal Welfare and Food Information for EU Consumers: Harmonizing the Regulatory Framework for More Policy Coherence. *European Journal of Risk Regulation* 1/2023.

[2] For instance, see Compassion in World Farming. EggTrack initiative: "Eggtrack." Available at: https://www.compassioninfoodbusiness.com/key-tools-for-success/eggtrack/.

[3] See, for instance, The Better Chicken Commitment. Available at: https://betterchicken-commitment.com/eu/

[4] See, for instance, The Accountability Board. Available at: https://accountability-board.org/.

[5] Marceau, J. 2019. *Beyond Cages: Animal Protection and Criminal Punishment*. Cambridge, Massachusetts: Cambridge University Press.

[6] For instance, see the legal action won by the Danish Vegetarian Association against Danish Crown, challenging the legality of climate-friendly claims on industrial farmed pork in 2024. Also see Chap. 11, Marine Friant-Perrot, "Meat Reduction as an Animal Protection Policy Goal: Exploring the Economic, Health, and Climate Nexus in EU Law."

[7] Block, K. and S. Amundson. 2025. We are asking the SEC to investigate the world's largest meat company's animal welfare promises. Humane World for Animals, 31 January. Available at: https://www.humaneworld.org/en/blog/jbs-pilgrims-pride-misleading-investors-animal-welfare?utm_source=chatgpt.com.

[8] For instance, Sykes, K. 2019. WTO law, the Environment and Animal Welfare. In *Animal Welfare and International Environmental Law: From Conservation to Compassion*, ed. Werner Scholtz, p. 270. Cheltenham: Edward Elgar.

to animals, successful claims under unfair commercial practices often result in significantly harsher penalties for defendants than those imposed in animal cruelty cases.[9] Furthermore, when market integrity is threatened by unlawful corporate behavior, legislatures often grant standing to a broad range of interested third parties, enabling private enforcement—a mechanism that remains largely absent from animal protection statutes.[10]

However, the growing use of economic law arguments and instruments in animal law-related litigation is not just the latest trend in animal advocacy. This turn to economic law is also of legislative and theoretical importance. On one hand, international organizations[11] and legislatures[12] increasingly integrate animal protection goals into corporate social responsibility standards. On the other, the shift to economic law to advance animal interests highlights potential overlaps between economic objectives and animal protection goals, while also raising critical questions about the possibility of transforming economic law in scenarios where such overlaps do not (yet) exist.

The EU provides a fertile ground for examining how legal frameworks can reconcile economic law's traditional focus on economic efficiency with the advancement of animal interests. EU animal welfare laws are indeed intertwined with economic law. This interconnection is often attributed to the EU's lack of an explicit competence to legislate on animal welfare. As a result, animal welfare measures have been developed under the umbrella of other policy areas where the EU does hold legislative authority—most notably the common agricultural policy, the "internal market" (i.e., the

[9] Tribunal correctionnel de Pau, *Régie des Abattoirs du Pays de Soule,* N° 16,090,000,004, 29 October 2019. Available at: https://www.l214.com/wp-content/uploads/2023/08/JUG-TA-PAU-210723-ABATTOIR-SOULE.pdf.

[10] Mo, D. 2005. Unhappy Cows and Unfair Competition: Using Unfair Competition Laws to Fight Farm Animal Abuse. *UCLA Law Review* 52, p. 1313 and 1315.

[11] See, for instance, OECD and UN Food and Agriculture Organization (FAO). 2016. Guidance for Responsible Agricultural Supply Chains. Available at: https://mneguidelines.oecd.org/oecd-fao-guidance.pdf. See also OECD. 2018. OECD Due Diligence Guidance for Responsible Supply Chains in the Garment and Footwear Sector. Available at: https://www.oecd.org/en/publications/oecd-due-diligence-guidance-for-responsible-supply-chains-in -the-garment-and-footwear-sector_9789264290587-en.html

[12] See Chap. 6, Clémentine Baldon, "Corporate Social Responsibility and Animal Protection."

EU market),[13] and the common commercial policy (i.e., trade with non-EU countries).[14]

While this dynamic has helped to mainstream animal welfare standards across the EU's fields of competence, EU animal welfare laws remain limited in scope and ambition. In this context, engaging with the legal structures of economic governance is necessary for scholars seeking to advance animal protection more effectively. Without a fundamental reorientation of economic law—one that reconfigures corporate incentives and aligns them with the interests of animals—meaningful progress remains unlikely.

This chapter begins by providing an overview of the economic foundations of EU animal welfare laws (Sect. 2). It then offers a critical analysis of economic law, exploring the potential opportunities for integrating animal protection standards into EU economic law (Sect. 3).

2 The Economic Foundations of Animal Protection Rules in EU Law

The EU legislature explicitly leverages its competence in regulating the internal market as a basis for animal welfare regulation. The EU legislature thus seems to consider animal welfare as a factor influencing market dynamics. This market-driven approach to animal welfare is not necessarily at odds with advancing higher levels of animal protection, as many legislative measures have been grounded in both market integration goals and ethical considerations (Sect. 2.1). In this context, the continued inadequacy of animal welfare standards in EU law (discussed in Sect. 2.4), despite the recognition of animal sentience in the EU constitutional treaty (Sect. 2.2), is more a result of the influence exerted by private dominant interest groups on legislative and rule-making processes than of a lack of explicit EU competence in the field of animal protection (Sect. 2.3).

2.1 The Dual Market-Based and Ethical Impetus for Animal Welfare Laws in the Pre-Lisbon Context

The constitutional provisions listing the fields and degree of competences of the EU do not list animal protection (or animal welfare) as a field of EU

[13] TFEU, art 114.

[14] Ryland, D. and A. Nurse. 2013. Mainstreaming after Lisbon: Advancing Animal Welfare in the EU Internal Market. *European Energy and Environmental Law Review*: 101–115.

competence.[15] However, the EU legislature has adopted a substantial number of animal protection rules by virtue of its competence in regulating economic activities that rely on animal exploitation. To this extent, EU animal welfare laws are primarily *economic* regulations. For instance, the EU leverages its broad competence in harmonizing legislation on the internal market through the adoption of common regulatory standards in EU law[16] to regulate the treatment of animals used for scientific purposes. Similarly, the EU draws on its competence in regulating agriculture[17] to govern the treatment of farmed animals, insofar as living animals are classified as agricultural products.[18] This allows the EU to adopt legislation aimed at achieving the economic objectives of its agricultural policy.[19] All trade bans benefiting animals, such as bans on the sale of cat fur and seal products, were also adopted on the basis of the EU's competence in regulating trade.[20]

As a result, EU farmed animal welfare legislation primarily aims to "eliminate distortions of competition" that "[interfere] with the smooth running of the organization of the market in animals" and to establish "common minimum standards for the protection of animals kept for farming purposes" so as to "ensure the rational development of production" on the EU market.[21] The prominence of competition objectives is evident

[15] Consolidated version of the Treaty on European Union, art 3 and 4, 2012 O.J. (C 326) 17.

[16] Ryland, D. and A. Nurse. 2013. Mainstreaming after Lisbon: Advancing Animal Welfare in the EU Internal Market. *European Energy and Environmental Law Review*, p. 114.

[17] TFEU, art 49.

[18] TFEU, art 38 and Annex I. In that sense, also see Sowery, K. 2018. Sentient Beings and Tradable Products: the Curious Constitutional Status of Animals Under Union Law. *Common Market Law Review*; 7–8; Ryland, D. and A. Nurse. 2013. Mainstreaming after Lisbon: Advancing Animal Welfare in the EU Internal Market. *European Energy and Environmental Law Review*;: 101–115; Di Concetto, A. 2024. EU Animal Law. In *Nordic Animal Law: Welfare and Rights*, ed. Birgitta Wahlberg. Cambridge: Ethics International Press.

[19] For more on the objectives of the Common Agricultural Policy, see also Chap. 3, Nicolas de Sadeleer and Alice Di Concetto, "The Common Agricultural Policy and Animals: Friends or Foes?"

[20] For instance, Regulation (EC) No 1007/2009 of the European Parliament and of the Council of 16 September 2009 on trade in seal products, 2009 O.J. (L 286) 36–39. For an exhaustive list of the competence used by the EU when legislating on animal protection issues, see Di Concetto, A. 2024. EU Animal Law. In *Nordic Animal Law: Welfare and Rights*, ed. Birgitta Wahlberg. Cambridge: Ethics International Press.

[21] Council Directive 98/58/EC of 20 July 1998 concerning the protection of animals kept for farming purposes, 1998 O.J. (L 221) 23–27.

in the recitals of EU farmed animal welfare laws, where the legislature presents fair competition as a primary rationale for regulatory intervention, while animal protection is framed as a secondary consideration.[22]

Similarly, the EU adopted a ban on the import and sale of seal products to correct "differences between national provisions governing the trade, import, production and marketing of seal products" that "adversely affect the operation of the internal market in products which contain or may contain seal products, and constitute barriers to trade in such products."[23] However, in subsequent challenges by the EU's trade partners before the World Trade Organization, the EU claimed to pursue public morals goals.[24]

The EU legislature also pursued economic goals through the adoption of a ban on the importation and sale of cat and dog fur, aiming to harmonize rules across Member States to ensure the "smooth operation of the internal market" and to prevent consumers from being misled.[25] Nevertheless, the legislature also recognizes that "cats and dogs are considered to be pet animals and therefore it is not acceptable to use their fur or products containing such fur."[26]

The Court of Justice of the EU (CJEU) validated the EU's indirect competence in the area of animal protection in a 1988 case, when the UK challenged the competence of the then-EU to enact rules on the welfare

[22] One exception is the Council Regulation (EC) 1/2005 on the protection of animals during transport and related operations and amending Directives 64/432/EEC and 93/119/EC and Regulation 1255/97 (Transport Regulation), 2004 O.J. (L 3) 1–44.

[23] Regulation (EC) No 1523/2007 of the European Parliament and of the Council of 11 December 2007 banning the placing on the market and the import to, or export from, the Community of cat and dog fur, and products containing such fur, Recitals 6 and 7, 2007 O.J. (L 343) 2. See also Di Concetto, A. 2024. EU Animal Law. In *Nordic Animal Law: Welfare and Rights*, ed. Birgitta Wahlberg. Cambridge: Ethics International Press.

[24] WTO. 2013. European Communities—Measures Prohibiting the Importation and Marketing of Seal Products. WT/DS400/R, WT/DS401/R; WTO. 2014. European Communities—Measures Prohibiting the Importation and Marketing of Seal Products. Appellate Body Report. WT/DS400/AB/R, WT/DS/401/AB/R.

[25] Regulation (EC) No 1523/2007 of the European Parliament and of the Council of 11 December 2007 banning the placing on the market and the import to, or export from, the Community of cat and dog fur, and products containing such fur, Recital 1, 2007 O.J. (L 343) 1. See also Di Concetto, A. 2024. EU Animal Law. In *Nordic Animal Law: Welfare and Rights*, ed. Birgitta Wahlberg. Cambridge: Ethics International Press.

[26] Regulation (EC) No 1523/2007 of the European Parliament and of the Council of 11 December 2007 banning the placing on the market and the import to, or export from, the Community of cat and dog fur, and products containing such fur, Recital 1, 2007 O.J. (L 343) 1.

of egg-laying hens. In an action for annulment, the UK argued that the EU did not have competence in regulating animal welfare matters. The Court ruled that the EU had not overstepped the limit of its powers because the adoption of rules concerning the welfare of egg-laying hens was based on the EU's competence to regulate the EU's agricultural market. Specifically, the Court decided that the EU has the competence to enact harmonizing animal welfare rules because "varying national rules regarding agricultural products [...] may affect the proper functioning of a common organization of the market, such as, in this case, differing conditions for the keeping of laying hens."[27] However, the Court also acknowledged that "the [Hens Directive] was also conceived with a view to ensuring better treatment for laying hens [...]."[28]

2.2 Post-Lisbon Animal Law: The Limited Effect of the Recognition of Animal Sentience in EU Constitutional Law

Consistent with the view that "the Union is not only a market to be regulated, but also has values to be expressed,"[29] the EU institutions amended the EU constitutional treaty in 2007 to include Article 13, Treaty on the Functioning of the European Union (TFEU), which states that:

> In formulating and implementing the Union's agriculture, fisheries, transport, internal market, research and technological development and space policies, the Union and the Member States shall, since animals are sentient beings, pay full regard to the welfare requirements of animals, while respecting the legislative or administrative provisions and customs of the Member States relating in particular to religious rites, cultural traditions and regional heritage.[30]

The EU thus became one of the few jurisdictions to codify the recognition of animal sentience in its constitution.[31] Analysts have questioned

[27] Case C-131/86, *United Kingdom of Great Britain and Northern Ireland v Council of the European Communities*, 23 February 1988, § 26–27.

[28] Ibid., § 27.

[29] Case C-34/10, *Oliver Brüstle v Greenpeace eV*, Judgment of the Court (Grand Chamber), 18 October 2011, Opinion of Advocate General Bot, 10 March 2011, § 46.

[30] TFEU, art 13.

[31] Eisen, J. and K. Stilt. Protection and Status of Animals. *Max Planck Encyclopedia of Comparative Constitutional Law [MPECCoL]*. Oxford: Oxford University Press.

how this modification of the Treaty might affect the substance of internal market legislation adopted by EU institutions.[32]

Such recognition remains limited in practice, as the treaty grants the EU no specific competence regarding animal protection, neither explicitly in Articles 3 and 4, nor implicitly through Article 13.[33] In the absence of a dedicated competence, the EU continues to regulate animal protection under the same legal bases used prior to the entry into force of Article 13. As a result, the objectives of EU animal protection rules have remained dual in nature: both economic and non-economic.

For instance, Directive 2010/63/EU on the Protection of Animals Used for Scientific Purposes[34] takes Article 114, TFEU on the approximation of laws (i.e., regulatory harmonization on the EU market) as a legal basis. With this Directive, the EU legislature explicitly pursues economic objectives, stating that "disparities between laws, regulations and administrative provisions of the Member States regarding the protection of animals used for experimental and other scientific purposes [...] are liable to constitute barriers to trade in products and substances the development of which involves experiments on animals. Accordingly, this Directive should provide for more detailed rules in order to reduce such disparities by approximating the rules applicable in that area and to ensure a proper functioning of the internal market."[35]

Economic objectives also feature prominently in the most recent reform proposals for new animal welfare regulations issued by the European Commission, which both take Article 43(2) (agricultural policy) and 114 (internal market) as legal bases.[36] More specifically, the proposal for a new

[32] De Witte, B. 2010. A Competence to Protect: The Pursuit of Non-Market Aims through Internal Market Legislation. In *The Judiciary, the Legislature and the EU Internal Market*, ed. Phil Syrpis, p. 44. Cambridge: Cambridge University Press.

[33] Bouhier, V. 2013. Le difficile développement des compétences de l'Union européenne dans le domaine du bien-être des animaux. *Revue Semestrielle de Droit Animalier* 1; Ryland, D. and A. Nurse. 2013. Mainstreaming after Lisbon: Advancing Animal Welfare in the EU Internal Market. *European Energy and Environmental Law Review*, p. 101 and 103.

[34] Directive 2010/63/EU on the protection of animals used for scientific purposes, 2010 O.J. (L 276) 33–79.

[35] Directive 2010/63/EU on the Protection of Animals Used for Scientific Purposes, Recital 1, 2010 O.J. (L 276) 33.

[36] Proposal for a Regulation of the European Parliament and of the Council on the Protection of Animals During Transport and Related Operations, Amending Council Regulation (Ec) No 1255/97 and Repealing Council Regulation (Ec) No 1/2005; European Commission. 2023. Proposal for a Regulation of the European Parliament and of the Council on the Welfare of Dogs and Cats and Their Traceability. Com/2023/769 Final.

Regulation on Dog and Cat Welfare and their Traceability—marking the EU's first regulatory intervention in this specific area—is primarily driven by competition-related concerns. These include disparities in animal welfare breeding standards and the sale of cats and dogs within the EU market, responding to "the sub-optimal functioning of the internal market for dogs and cats in the Union as well as of illegal trade in these animals within the Union and at import into the Union."[37]

Similarly, the proposed Regulation on the Transport of Animals seeks to address key shortcomings of current rules—particularly regarding transport times[38] and exports to third countries—primarily to "ensure the smooth functioning of the internal market."[39] Only a handful of the new proposed rules would improve the welfare of animals, while some may even worsen it, such as the proposed extension of transport times for pigs by boat, which the European Commission introduced to align with transport times for other mammals, to standardize rules across species.[40]

In this context, some animal legal scholars and advocates have attributed the weak level of animal protection in EU law to the absence of an explicit competence in this area.[41] Establishing a specific competence regarding animal welfare could create an obligation for the legislature to enact new, additional, and more comprehensive standards, as well as to integrate animal welfare considerations beyond the limited set of policies listed in Article 13, TFEU. Creating an explicit legal basis related to animal protection in EU law further seems essential to eliminate the

[37] European Commission. 2023. Proposal for a Regulation of the European Parliament and of the Council on the Welfare of Dogs and Cats and Their Traceability. Com/2023/769 Final, Recital 1.

[38] The rules on transport time in current legislation are so unclear that economic operators brought many challenges before the Court of Justice of the EU in an attempt to clarify them. See Di Concetto, A. 2022. For a More Humane Union: Legal Assessment of EU Farm Animal Welfare Law. *The European Institute for Animal Law & Policy*.

[39] Proposal for a Regulation of the European Parliament and of the Council on the Protection of Animals During Transport and Related Operations, Amending Council Regulation (Ec) No 1255/97 and Repealing Council Regulation (Ec) No 1/2005, Recital 4.

[40] Poinsot, A. Forthcoming 2026. La protection des animaux d'élevage lors du transport routier à l'épreuve de la réforme du droit de l'Union européenne. In *Regards croisés: le législateur face aux animaux d'élevage*, ed. Adrienne Bonnet. LGDJ-Lextenso.

[41] For instance, Bouhier, V. 2013. Le difficile développement des compétences de l'Union européenne dans le domaine du bien-être des animaux. *Revue Semestrielle de Droit Animalier* 1.

commodification of animals[42] and pave the way to a change in the legal status of animals.

From a more practical perspective, however, it seems that the absence of such a legal basis is not an obstacle to the adoption of ambitious animal protection rules. The CJEU's approach to laws pursuing economic and ethical objectives has been notably favorable to ethical objectives: "as long as the measure improves the conditions for the establishment and functioning of the internal market, it is an appropriate legal basis, even if non-economic considerations were decisive for its adoption."[43] Consequently, "there are no major legal obstacles that prevent the inclusion of non-market values in internal market legislation."[44]

For instance, while the EU legislature regulates the welfare of egg-laying hens in the egg industry to harmonize egg production standards in the EU, the legislature has done so in a way that would prohibit the use of unenriched cages, which slightly improves animal welfare practices compared to common industry practices.[45] Consistent with this logic, in the 1986 court case that opposed the United Kingdom to the then-European Commission, the CJEU admitted that "the [Directive] was also conceived with a view to ensuring better treatment for laying hens," but did not find the Directive to be contradicting its primary objective, which was to harmonize the "differing conditions for the keeping of laying hens."[46]

On another note, the EU legislature has, in a few notable instances, enacted animal protection measures that fall outside the strict boundaries

[42] Rules on animal protection grounded in the EU's competence to regulate the internal market can pursue animal welfare objectives but it cannot be their exclusive goal. "There is always a threshold requirement that the measure must also adequately contribute to improving the conditions for the establishment and functioning of the internal market." Source: De Witte, B. 2010. A Competence to Protect: The Pursuit of Non-Market Aims through Internal Market Legislation. In *The Judiciary, the Legislature and the EU Internal Market*, ed. Phil Syrpis, p. 36, note 32. Cambridge: Cambridge University Press.

[43] Kosta, V. 2015. *Fundamental Rights in EU Internal Market Legislation*, p. 22. Oxford: Hart Publishing.

[44] De Witte, B. 2010. A Competence to Protect: The Pursuit of Non-Market Aims through Internal Market Legislation. In *The Judiciary, the Legislature and the EU Internal Market*, ed. Phil Syrpis, 25–46. Cambridge: Cambridge University Press.

[45] Council Directive 1999/74/EC of 19 July 1999 laying down minimum standards for the protection of laying hens, 1999 O.J. (L 203) 53–57.

[46] C-131/86, *United Kingdom of Great Britain and Northern Ireland v. Council of the European Communities*, 1988 ECLI:EU:C:1988:86, § 26.

of its formal competences.[47] One example is the adoption of animal welfare rules in areas not listed under Article 13—such as trade—indicating that an autonomous EU competence on animal welfare may not be necessary for the Union to extend its actions beyond the material scope of Article 13.

Rather, the resistance to adopting more ambitious animal welfare laws stems from political dynamics at EU level,[48] which led the EU to defend, for the past three decades at least, the goal of enhancing the EU's exporting power on the global market. This vision conflicts with the goal of raising production standards, as such standards could place EU producers at a competitive disadvantage unless trade partners agree to adopt equivalent standards, which could complicate trade negotiations.

Similarly, national legislatures are often not able to adopt stricter national animal welfare rules due to their inability to erect trade barriers within the EU and protect their domestic markets from crueler products originating from other Member States.[49] As a consequence, "unilateral moves to protect farmed animal welfare on the part of individual Member States" are unlikely, since they "would put their national industry at a competitive disadvantage."[50]

2.3 *The Influence of Dominant Private Interest Groups on the Legislative Process and Rule-Making Procedures*

Rather than the absence of a legal basis for animal welfare in the EU constitutional treaties, the lack of ambition in EU animal welfare laws is

[47] Similarly, the EU has legislated in the area of public health despite the absence of any explicit competence in this field. See Delhomme, V.N. 2024. Minimum Harmonization, Experimentation and the Internal Market: Balancing Unity and Diversity in the Regulation of Lifestyle-Related Health Risks. In *Balancing Unity and Diversity in EU Legislation*, ed. Ton van den Brink and Virginia Passalacqua, 194–210. Cheltenham: Edward Elgar.

[48] De Witte, B. 2010. A Competence to Protect: The Pursuit of Non-Market Aims through Internal Market Legislation. In *The Judiciary, the Legislature and the EU Internal Market*, ed. Phil Syrpis, p. 37, note 32. Cambridge: Cambridge University Press.

[49] As per Case C-5/94, *The Queen v. Ministry of Agriculture, Fisheries and Food, ex parte: Hedley Lomas (Ireland) Ltd.*, 1996 ECR I-02553, and Case C-1/96, *The Queen v Minister of Agriculture, Fisheries and Food, ex parte Compassion in World Farming Ltd.*,1996 ECR I-01251.

[50] Di Concetto, A. 2022. The Double Edged Sword: International Law and Its Effects on EU Farm Animal Welfare Legislation. *Global Journal of Animal Law, Special Issue: Internal Law and Animal Health and Protection: Persistent Themes, New Prospects for Change* 10(2).

primarily driven by the influence of certain vested interests.[51] Large corporations that use animals for commercial purposes have the resources to influence the legislative process, through the hiring of professional lobbyists for instance, but they also benefit from an institutional ecosystem that is prone to industry capture.[52]

The composition rules of the "Civil Dialogue Groups" (CDGs) are an example of how the EU institutions tend to favor the views of industry players and the most moderate voices in civil society organizations. CDGs serve as an official, formal mechanism of the European Commission for engaging with a wide range of civil society representatives in the agri-food sector, including farmers' unions, advocacy groups, and think tanks. At their inception in 1962, the EU executive branch created the CDGs to ensure a continuous dialogue between EU policymakers and farmers. In practice, CDGs are advisory groups sub-divided into seven sub-thematic groups, which are each co-chaired by representatives from the private sector and European Commission representatives.[53] Participants are appointed by the European Commission for a five-year mandate, following a call for application, and CDGs meet regularly to discuss agenda points decided collectively.

Rules related to the composition and governance of CDGs were reformed in 2004,[54] and again in 2013,[55] to include representation of all civil society stakeholders, including public interest groups.[56] More specifically, the 2013 reform mandated that the European Commission "ensure

[51] In that sense, see Scully, M. 2003. *Dominion: The Power of Man, the Suffering of Animals, and the Call to Mercy*. New York: St Martin's Griffin; Connor Cleveland, J. 2016. Why Animal Welfare Is a Conservative Cause. *National Review*, 19 April.

[52] Di Concetto, A. 2023. Le pouvoir des lobbies privés face au mouvement de protection animale: étude de cas au niveau européen. In *Les mutations contemporaines du droit de l'animal*, ed. Olivier Le Bot. Aix-en-Provence: Revue Confluence des droits (in French).

[53] Commission Decision (EU) 2022/1368 of 3 August 2022 setting up Civil Dialogue Groups in matters covered by the common agricultural policy and repealing Decision 2013/767/EU, art 1, 2022 O.J. (L 205) 279.

[54] Commission Decision 2004/391/EC of 23 April 2004 on the advisory groups dealing with matters covered by the common agricultural policy, 2004 O.J. (L 120) 50–60.

[55] Commission Decision 2013/767/UE of 16 December 2013 setting up a framework for civil dialogue in matters covered by the common agricultural policy and repealing Decision 2004/391/EC, 2013 O.J. (L 338) 115–117.

[56] European Commission. 2019. Study on the Civil Dialogue Groups for the Common Agricultural Policy—Analysis of EU Policy Consultation. Written by Deloitte, Areté SRL, AFC Consulting Group AG, and AGRA CEAS Consulting SA.

a balanced representation of all expressed interests [...]. In particular, [...] a balance between economic and non-economic interests."[57] However, the 2013 reform did not remedy the fact that CDG member organizations were afforded a number of seats in proportion to their membership base, thus leading to a disproportionate amount of seats for dominant private interest groups compared to public interest groups.

Additionally, the 2013 reform required that CDG members "be composed of at least European-level non-governmental organizations."[58] This criterion benefited umbrella organizations and federations, thereby excluding smaller organizations who consciously choose not to invest time and resources in membership fees, opting to conduct EU policy work independently. Another limitation of granting representation exclusively to federations concerns the positions adopted by these federations, which can become diluted by the diverse goals of their members. This often leads federations to adopt compromise positions that are inconsistent with their base.

The European Commission further reformed CDG rules in 2022, partly remedying the disproportionate representation of private interest groups in CDGs and opening membership applications to all organizations registered in the EU Transparency Register. In practice, however, the composition of CDGs has remained virtually unchanged compared to 2013.[59]

The disproportionate representation of dominant private interest groups is problematic for at least three reasons. First, it goes against the interests of animals. Second, such unfair representation in law- and rule-making procedures is not consistent with the democratic values the EU claims to uphold. Third, the primacy of dominant economic interests over animal welfare objectives undermines economic efficiency. Many of the tepid animal welfare rules the EU has enacted over the past four years under the influence of vested interests have imposed significant costs on

[57] Commission Decision 2013/767/UE of 16 December 2013 setting up a framework for civil dialogue in matters covered by the common agricultural policy and repealing Decision 2004/391/EC, art 4(3), 2013 O.J. (L 338) 116.

[58] Ibid., art 4(2).

[59] The composition of the membership of all the CDGs is available online on the European Commission's website. For instance, the CDGs on "animal products" are available at: https://ec.europa.eu/transparency/expert-groups-register/screen/expert-groups/consult?lang=en&groupID=3879.

economic operators, thereby reinforcing market consolidation, limiting market access and weakening the market's self-regulatory powers.

The European Commission itself admitted in its 2022 evaluation of EU farm animal welfare laws that "[t]he EU animal welfare legislation has contributed to, but not fully ensured, equal conditions for the operators and the economic activities affected. Differences in application and enforcement still create obstacles to the internal market and the achievement of comparable levels of animal welfare."[60] In other words, the Commission also deplores the ineffectiveness of EU animal welfare laws in achieving both competitive efficiency and meaningful animal protection.

2.4 *The Resulting Limitations of EU Animal Welfare Laws*

The influence of powerful corporate actors on animal welfare laws is evident in several characteristics of EU animal welfare legislation that undermine its effectiveness.

A first shortcoming of EU animal welfare laws is that many of the standards contained in them are drafted using general terms to such an extent that they leave a significant margin for interpretation by Member States, who often favor the views of dominant economic operators.[61] As a consequence, these laws have produced little to no effect on the levels of protection afforded to animals. EU farmed animal welfare legislation typically uses open-ended standards, such as the need to avoid "unnecessary suffering," or acts "likely to cause suffering or injury," and conversely, the requirement to offer "appropriate," "adequate," or "comfortable" treatment.[62] However, these open-ended, subjective standards are left largely unsubstantiated.[63] Another example of such imprecision is in the wording of Directive 2010/63 regarding the composition and governance of ethics

[60] European Commission. 2022. Commission Staff Working Document Fitness Check of the EU Animal Welfare legislation. SWD(2022) 329 final, p. 62.

[61] In that sense, see the 2024 ruling by a Dutch court in *Stichting Varkens in Nood te Amsterdam (ViN) en de minister van Landbouw, Visserij, Voedselzekerheid en Natuur*, 2024 ECLI:NL:CBB:2024:578. The ruling found that the Dutch had insufficiently motivated the legality of the use of farrowing crates for sows in Dutch national legislation.

[62] Directive 98/58 of 20 July 1998 concerning the protection of animals kept for farming purposes, Annex, 1998 O.J. (L 221) 26–27.

[63] For a more detailed analysis of the many shortcomings of EU farm animal welfare legislation, see Di Concetto, A. 2022. For a More Humane Union: A Legal Assessment of EU Farm Animal Welfare Laws. *The European Institute for Animal Law & Policy.*

committees responsible for evaluating and approving animal-based research. As a result, significantly different practices take place across the Union when it comes to the use of animals for scientific purposes. These disparate practices often end up harming both fair competition and animal protection objectives.[64]

A second shortcoming of EU animal welfare laws lies in the fact that large portions of animals are in fact excluded from the scope of key legislative acts. Such is the case for fish, who are covered under Directive 98/58 as vertebrate animals, as well as the Transport Regulation. However, none of the provisions in these two acts apply to fish, since the EU legislature primarily intended the rules to apply to terrestrial animals. Fish are further excluded from almost all specific provisions in the Slaughter Regulation.[65] Similarly, many farmed animals are not covered by species-specific directives. Consequently, their protection is effectively absent, due to the lack of targeted provisions in Directive 98/58. This is the case for geese and ducks, for instance, who are exposed to cruel industry practices in the context of foie gras production. Similarly, dairy and beef cows only receive protection under the Calves Directive[66] until they are six months of age. Lastly, the vast majority of EU farm animal welfare standards do not apply to animals and animal-source food imported into the EU.[67]

A third shortcoming of EU animal welfare laws is that exemptions are also commonplace. For instance, the Broiler Directive sets a maximum density rate of 33 kg/m^2 on broiler chicken farms, all the while allowing higher density levels "by way of derogation" provided producers comply with additional requirements, including the monitoring of the number of animals dead on arrival at the slaughterhouse.[68] Sixty percent of EU broiler

[64] Di Concetto, A. and A. Poinsot. 2024. L'évaluation éthique des projets de recherche impliquant l'utilisation d'animaux: étude de droit comparé dans l'Union européenne. *Bulletin de l'Académie vétérinaire de France* (in French).

[65] Council Regulation 1099/2009 of 24 September 2009 on the Protection of Animals at the Time of Killing, art 1, O.J. (L 303) 7.

[66] Council Directive 2008/119 of 18 December 2008 Laying Down Minimum Standards for the Protection of Calves, art 2(1), 2009 O.J. (L 10) 8, in light of C-355/11, *G. Brouwer Versus Staatssecretaris van Economische Zaken, Landbouw en Innovatie*, 2012 ECLI:EU:C:2012:353.

[67] The Slaughter Regulation applies extraterritorially, for sanitary purposes. The Pigs and Calves Directives only apply to live animals imported to the EU.

[68] Directive 2007/43 of 28 June 2007 Laying Down Minimum Rules for the Protection of Chickens Kept for Meat Production, art 3, Annex III, and V, 2007 O.J. (L 182) 21, 26, and 27.

producers make use of this derogation, which has become more of a common industry practice than a derogation.[69]

At a more structural level, a fourth shortcoming of EU farm animal welfare laws and animal welfare laws in general, is that they operate in isolation from the legal frameworks governing corporations and markets. The EU legislature incentivizes market actors to prioritize short-term financial returns, and treats animal welfare as an isolated concern, detached from the criteria used to evaluate corporate performance and managerial effectiveness. The current framework thus fundamentally encourages corporate resistance to efforts aimed at improving animal welfare.

3 Integrating Animal Protection Goals Into Economic Law

Animal rights advocates and animal law experts often view economic law as a barrier to advancing more ambitious animal protection standards. While this perception has merit, a more nuanced approach is needed to avoid essentializing certain trends in the evolution of economic law and to recognize its potential for reassessment and reform. The current context appears favorable for such a reevaluation, as the EU has shown, in recent years, growing interest in integrating sustainability and animal welfare goals into economic law. This shift should encourage legal scholarship to move beyond traditional "dichotomous thinking,"[70] and stop assuming that economic law is exclusively concerned with narrow economic interests, leaving other legal domains responsible for addressing the negative consequences of corporate activities and market transactions (3.1). Indeed, the insufficient integration of animal protection goals into economic law instruments may largely explain the stagnation in animal legislative developments positioning economic law as a critical foundation for the transition toward greater consideration for animals (3.2).[71]

[69] European Commission. 2017. Report on the Application of Directive 2007/43/EC and Development of Welfare Indicators, p. 8.

[70] Offor, I. 2020. Animals and the Impact of Trade Law and Policy: A Global Animal Law Question. *Transnational Environmental Law* 9(2): 239–262.

[71] On the concept of "consideration" and its added value compared to other notions such as "protection," see notably Pelluchon, C. 2018. *Ethique de la considération*. Paris: Editions du Seuil (in French).

3.1 Economic Law's Indifference to the Interests of Animals Grounded in a False Dichotomy Between Economic Law and Animal Law

Economic law is often perceived as inherently indifferent to animal protection, a stance that has faced significant criticism. Far from being neutral, this neglect of animal interests has been argued to have "done more harm than good to animals."[72] Traditionally, critics of economic law's neglect of animals have advocated for the development of a separate legal domain, exclusively focused on advancing animal interests, designed to override market-driven logics and the rules that sustain them (Sect. 3.1.1.). In recent years, however, another perspective has gained prominence, one that challenges the foundational binary between economic law and so-called external regulations aimed at protecting human rights, the environment, or animals (Sect. 3.1.2.).[73]

3.1.1 Rethinking the Binary Between Economic Law and "External Regulations"

The constructed binary between economic and non-economic fields of law can be traced back to the doctrinal efforts of certain prominent Western legal scholars in the 1970s. These scholars sought to assign strictly economic purposes to areas of law categorized as "economic." Moreover, they offered a narrowly defined understanding of these "strictly economic purposes." Legal rules were deemed economic—and thereby fell within the scope and mandate of economic rules and institutions—only if they were designed to maintain the perceived internal dynamics of the economic system. This included ensuring the proper and supposedly natural functioning of markets or businesses, with a particular focus on promoting free and perfectly competitive markets. All other rules were classified as political in nature, including those that exert significant influence on companies and markets but whose primary regulatory purpose is to protect the

[72] Peters, A. 2021. *Animals in International Law*. Leiden: Brill Nijhoff, p. 535.

[73] See notably Epstein, A.-S., coord. 2025. La Transformation écologique du droit économique. Report of the TEDE Research Group (in French). Available at: https://ecologic.law/publications-2/la-transformation-ecologique-du-droit-economique/; Viñuales, J. E. and J.-F. Mercure. 2020. Pathway to Reframing Environmental Law. *Environmental Policy and Law* 50, p. 509; Sjåfjell, B. and C. M. Bruner, eds. 2020. *The Cambridge Handbook of Corporate Law, Corporate Governance and Sustainability*. Cambridge: Cambridge University Press; Bartl, M. 2024. *Reimagining Prosperity: Toward a New Imaginary of Law and Political Economy in the EU*. Cambridge: Cambridge University Press.

interests of parties who are not considered economic agents involved in market transactions. The distinction between international *trade* agreements, covered by international trade law, and *environmental* treaties, such as the Convention on International Trade in Endangered Species, was drawn along these lines.[74]

This division between economic and political rules reflects the influence of neoclassical microeconomics on law and legal thought, particularly its distinction between transactions—regarded as strictly economic operations falling under the purview of economic law—and externalities, which were relegated to the realm of "external regulations."[75] Such a binary framework can also be seen as an attempt to shield economic agents—not only businesses but also economic regulators—from government intervention.[76]

Proponents of the binary distinction between economic law and external regulations often present it as a means to enhance economic efficiency, including by fostering the rationality of economic decision-making. By assigning legal domains deemed "economic" a single and narrowly defined focus, this approach is presumed to yield higher efficiency. This claim is premised on the principle that "[i]t is logically impossible to maximize in more than one dimension at the same time."[77] This means that economic decision-makers whether from the public sector (e.g., competition authorities)[78] or the private sector (e.g., corporate managers),[79] should concentrate on maximizing a single narrow objective (e.g., consumer welfare, shareholder value, financial stability), as opposed to balancing competing interests and objectives. This one-dimensional approach is thus said to protect companies and markets from the perceived arbitrariness of eco-

[74] See more broadly Couveinhes Matsumoto, F. 2023. *Droit International Économique*. Paris: Presses Universitaires de France (in French).

[75] Viñuales, J. E. and J.-F. Mercure. 2020. Pathway to Reframing Environmental Law. *Environmental Policy and Law* 50, p. 509.

[76] See, for instance, Friedman, M. 1970. The Social Responsibility of Business is to Increase its Profits. *The New York Times Magazine*, 13 September; Tirole, J. 2023. Socially Responsible Agencies. *Competition Law & Policy Debate* 7: 171–177.

[77] Jensen, M. C. 2001. Value Maximization, Stakeholder Theory, and the Corporate Objective Function. *Journal of Applied Corporate Finance* 14(3), p. 8 and 10.

[78] See, for instance, Petit, N. and L. Radic. 2024. The Superiority of the Consumer Welfare Standard. EUI Working Paper LAW 2024/20. Available at: https://papers.ssrn.com/sol3/papers.cfm?abstract_id=5065469; Tirole, J. 2023. Socially Responsible Agencies. Competition Law & Policy Debate 7, note 77.

[79] Bebchuk, L. A. and R. Tallarita. 2020. The Illusory Promise of Stakeholder Governance. *Cornell Law Review* 106: 91–177.

nomic decision-makers—arbitrariness that is thought to arise when these actors are required to weigh conflicting considerations.

Since the 1970s, the dichotomy between "economic" rules and "non-economic" regulations has permeated economic law worldwide, giving rise to a series of distinctions aimed at narrowing down the scope and mandate of economic rules and institutions. For instance, while "[a]nimal welfare is being increasingly perceived as an integral element of overall food quality,"[80] international trade experts have strongly differentiated *products* and production *processes*, thereby limiting the ability of governments to promote a shift toward more humane production methods. Based on this distinction, subjecting different *products* to different rules is not considered inherently discriminatory or illegal. However, there has been consistent reluctance to legitimize the application of a differentiated legal regime to products based solely on their production *processes*.[81] This reluctance stems from the belief that differences in treatment based on production methods could serve as an easy gateway for governments to introduce irrational and economically suboptimal protectionist measures, disguised as efforts to promote animal welfare or other non-market objectives.

3.1.2 Critics of the Alleged Binary Between Economic Law and Animal Law

The exclusion of animal protection objectives from the domain of economic law is highly contestable. The purported binary between economic regulation and animal welfare regulations not only contributes to the continued harm of animals (Sect. 3.1.2.1.), but also rests on problematic and theoretically unsound foundations (Sect. 3.1.2.2.).

3.1.2.1 A Cause of Harm to Animals

The absence of animal protection consideration within economic law results in significant harm to animals—harm that may be difficult to quantify, yet is well documented through numerous concrete examples.

Scholars have been particularly critical of the contradictions between animal protection objectives and the frameworks of international trade law

[80] Horgan, R. and A. Gavinelli. 2006. The Expanding Role of Animal Welfare within EU Legislation and Beyond. *Livestock Science* 103(3): 303–307.

[81] For a critical appraisal, see notably Kysar, D.A. 2004. Preferences for Processes: The Process/Product Distinction and the Regulation of Consumer Choice. *Harvard Law Review* 118(2): 526–642.

and investment treaties. The liberalization of international trade flows has facilitated the long-distance transport of live animals and accelerated the segmentation and industrialization of animal agriculture.[82] Likewise, the neglect of animal welfare by international lending institutions, and within international investment agreements, has drawn scrutiny, as it has significantly contributed to the global proliferation of factory farming.[83]

These developments have not been counterbalanced by more ambitious animal protection policies. On the contrary, international trade law has been criticized for exerting a chilling effect on the readiness of governments to enact stronger animal welfare statutes or even implement mandatory labeling schemes that would shed some light on production methods. For instance, the EU does not prohibit the importation of animal products that fail to meet animal welfare standards equivalent to those required within its own territory. As a result, the EU continues to import food products—such as eggs and veal—from farming systems that are prohibited within its own jurisdiction.[84] Additionally, the EU has refrained from

[82] See, for instance, FAO Committee on Agriculture. 2005. The Globalizing Livestock Sector: Impact of Changing Markets, 19th session. Available at: https://www.fao.org/4/j4196e/j4196e.htm?utm_source=chatgpt.com. See more broadly Nielsen, L. 2007. *WTO, Animals and PPM's*. Leiden: Brill Nijhoff; Kelch, T.G. 2011. *Globalization and Animal Law: Comparative Law, International Law and International Trade*. Alphen aan den Rijn: Kluwer Law International; Blattner, C. 2019. *Protecting Animals Within and Across Borders: Extraterritorial Jurisdiction and the Challenges of Globalization*. Oxford: Oxford University Press; Sykes, K. 2021. *Animal Welfare and International Trade Law: The Impact of the WTO Seal Case*. Cheltenham: Edward Elgar; Offor, I. 2024. *Global Animal Law From the Margins: International Trade in Animals and their Bodies*. London: Routledge.

[83] See notably Nollkaemper, A. 2023. International Law and the Agony of Animals in Industrial Meat Production. *The European Journal of International Law* 34(4), p. 945–946; McNamar, K. and K. Hamerschlag. 2023. Multilateral Banks' Investments in Industrial Livestock Undermine their Paris Climate Commitments. *Climate Home News*, 21 June. Available at: https://www.climatechangenews.com/2023/06/21/multilateral-banks-investments-in-industrial-livestock-undermine-their-paris-climate-commitments/; Columbia Center on Sustainable Development, International Institute for Environment and Development and International Institute for Sustainable Development. 2018. Agricultural Investments under International Investment Law. Available at: https://ccsi.columbia.edu/sites/default/files/content/docs/publications/CCSI-IIED-IISD_Agricultural-Investments-under-IIL.pdf.

[84] Offor, I. 2020. Animals and the Impact of Trade Law and Policy: A Global Animal Law Question. *Transnational Environmental Law* 9(2): 239–262. According to Offor, "The EU unilaterally restricts trade in some controversial animal products—such as seal, or cat and dog fur—but it permits and promotes trade in most animal products without conditioning this on animal welfare protection."

enforcing its ban on fur imports derived from the use of cruel leghold traps, due to concerns that this ban might be challenged by the United States or Canada before the World Trade Organization (WTO).[85]

While WTO rules do not inherently oppose animal welfare laws, the WTO typically categorizes animal welfare as a "non-trade" issue, placing it outside the organization's main trade mandate. Consequently, the WTO permits trade barriers based on animal welfare objectives only on a case-by-case basis. Furthermore, the WTO frames such barriers as exceptions, justifiable through anthropocentric motives such as public morals. This framing shifts the burden of proof for the legality of the measure onto the enacting State and invites narrow and unpredictable interpretations.[86]

Similarly, the objectives of competition law, such as enhancing consumer welfare, can at times come at the expense of animal welfare. For instance, in the 2015 *Chicken of Tomorrow* case,[87] the Dutch Competition Authority had to assess the legality of a sector-wide agreement which required participating companies to improve chicken welfare standards. This agreement was expected to raise the market penetration of higher-welfare poultry products from 9% to 100%,[88] which would have benefited chickens. However, the Dutch Competition Authority ultimately found the agreement to be unlawful, considering that consumers would be unwilling to pay the higher price agreed upon by three parties.

At the time, the European Commission was more generally prone to consider sustainability issues as falling outside the scope of competition law. For instance, in 2017–2018, the Commission rejected numerous

[85] Ibid., p. 254–256. See also Stevenson, P. 2015. The Impact of the World Trade Organisation Rules on Animal Welfare. Compassion in world farming. Available at: https://www.ciwf.org.uk/media/7131790/the-impact-of-the-world-trade-organisation-rules-on-animal-welfare-2015.pdf

[86] See, for instance, Kruse, R. 2013. Process and Production Methods and Burden of Proof: A Procedural Limitation on the Like Products Debate. *International Trade and Business Law Review* 16: 377–396; Cima, E. 2022. *From Exception to Promotion: Re-Thinking the Relationship between International Trade and Environmental Law*. Leiden: Brill Nijhoff.

[87] ACM's Analysis of the Sustainability Arrangements Concerning the 'Chicken of Tomorrow.' 2015. Case number: 13.0195.66. Available at: https://www.acm.nl/en/publications/publication/13761/Industry-wide-arrangements-for-the-so-called-Chicken-of-Tomorrow-restrict-competition. On this case, see notably Bos, J. M., van den Belt, H., and P. H. Feindt. 2018. Animal Welfare, Consumer Welfare, and Competition Law: The Dutch Debate on the Chicken of Tomorrow. *Animal Frontiers* 8(1): 20–26.

[88] Lelieveldt, H. 2020. Out of Tune or Well-Tempered? How Competition Agencies Direct the Orchestrating State. *Regulation & Governance* 14: 465–480.

petitions opposing the merger between the companies Bayer and Monsanto. Opponents of the merger claimed that a Bayer-Monsanto conglomerate would threaten food safety, consumer protection, and environmental preservation, including biodiversity conservation and climate targets. However, the Commission considered only narrow economic parameters when assessing the legality of the merger, stating that "the Commission's mandate under the European merger control rules is to assess the merger solely from a competition perspective." While the Commission recognized that "(…) [w]hen it comes to seeds and pesticides there are of course other vital concerns that go beyond competition policy, including consumer protection, food safety and ensuring the highest standards for the environment and the climate,"[89] it refused to recognize its competence in addressing the merger's impacts on public health, biodiversity, and consumers.[90]

3.1.2.2 Flawed Theoretical Justifications

Critics of economic institutions' inadequate consideration of social, environmental, or animal interests have often accepted the notion that economic law is inherently distinct from social, environmental, or animal protection policies.[91] In doing so, proponents of "trade and…" or "competition and…" scholarship inadvertently reinforced a binary view of the legal system where economic institutions are perceived as insulated from other public policy objectives.[92] However, the foundations underpinning the binary distinction between economic law and external regulations are conceptually weak.

[89] European Commission. 2018. Press Release, Mergers: Commission clears Bayer's acquisition of Monsanto, subject to conditions. Available at: https://ec.europa.eu/commission/presscorner/detail/en/ip_18_2282

[90] The Commission's approach to sustainability issues has evolved in the last few years, as exemplified by its Guidelines on sustainability agreements (e.g., the Guidelines on the applicability of Article 101 of the Treaty on the Functioning of the European Union to horizontal co-operation agreements, 2023 O.J. (C 259) 1–125), particularly in the agricultural sector (see the Commission guidelines on the exclusion from Article 101 of the Treaty on the Functioning of the European Union for sustainability agreements of agricultural producers pursuant to Article 210a of Regulation (EU) No 1308/2013, 2023 O.J. C/2023/1446).

[91] See, for instance, Descola, P. and A. Pignocchi. 2022. *Ethnographies des mondes à venir*, 82 sq. Paris: Seuil.

[92] Lang, A. T.F. 2007. Reflecting on "Linkage": Cognitive and Institutional Change in the International Trading System. *Modern Law Review* 70(4): 523–549, p. 536–537.

First, the coherence of the legal system depends on the absence of rigid, hermetically sealed divisions between different areas of law. As demonstrated in Part 2, economic rules frequently pursue non-market objectives, just as animal protection statutes often incorporate economic considerations. No area of law exists in isolation or in a state of "chemical purity." As a result, it would be "wrong to present internal market legislation as an efficiency-driven enclave amidst a multiplicity of value-driven EU policies."[93] The fact that "[p]olitical contestation about the normative preferences which the EU should pursue and promote increasingly takes place outside the domain of internal market legislation,"[94] in areas such as environmental or animal welfare policy, does not mean that animal protection should be excluded from the objectives of economic law.[95]

Second, restricting the scope of economic law to narrowly defined economic interests does not make it politically neutral, contrary to the assumptions underlying the traditional divide between economic law and so-called external regulations. On the contrary, this framing carries clear distributive consequences—often to the benefit of dominant economic actors.

The European Central Bank's (ECB) corporate bond purchasing program offers a clear example of how the purported "neutrality" of economic law tends to favor dominant players. Until 2021, the ECB's claim to political neutrality led it to buy corporate bonds in proportion to their presence in the market—the ECB was "buying the market." This approach led the bank to buy bonds from big agricultural firms, disregarding their

[93] De Witte, B. 2010. A Competence to protect: The Pursuit of Non-Market Aims through Internal Market Legislation. In *The Judiciary, the Legislature and the EU Internal Market*, ed. Phil Syrpis, p. 44. Cambridge: Cambridge University Press.

[94] Ibid. See also Lang, A. T.F. 2007. Reflecting on "Linkage": Cognitive and Institutional Change in the International Trading System. *Modern Law Review* 70(4): 523–549. Even authors who emphasize the potential of economic law and institutions to advance animal interests often view such considerations as "non-economic" or "non-market" (e.g., De Witte, B. 2010. A Competence to protect. The pursuit of non-market aims through internal market legislation. In *The Judiciary, the Legislature and the EU Internal Market*, ed. Phil Syrpis, p. 26–27. Cambridge: Cambridge University Press) and as naturally falling outside the mandate of economic institutions. See, for instance, Sykes, K. 2019. WTO law, the animal welfare. In *Animal Welfare and International Environmental Law: From Conservation to Compassion*, ed. Werner Scholtz, 269–304, p. 270. Cheltenham: Edward Elgar.

[95] Lang, A. T.F. 2007. Reflecting on "Linkage": Cognitive and Institutional Change in the International Trading System. *Modern Law Review* 70(4): 523–549, note 93.

impact on climate change or animal welfare.[96] The ECB hailed this approach as a means of avoiding the imposition of any particular conception of the public good, thereby shielding the market from potentially inefficient political interference. However, such private sector bond purchase programs were not politically neutral, as they disproportionately enhanced the financing capacity of large, established companies, often to the detriment of their smaller competitors. Given that the largest greenhouse gas emitters are typically overrepresented among large-cap stock companies, the ECB's private sector bond purchase programs have further reinforced unsustainable business models and, more broadly, have "[*reinforced*] *pre-existing inefficiencies that give rise to a suboptimal allocation of resources.*"[97]

Competition law and international trade law also illustrate the distributive consequences of economic law and its inherent lack of political neutrality. To differentiate between permissible government or corporate policies and unacceptable market distortions, both WTO law and competition law primarily rely on consumer-related standards—consumer welfare in competition law[98] and consumers' tastes and preferences in WTO law.[99] However, these concepts are interpreted based on buyers' *current* purchasing decisions. While this approach provides an empirical footing to economic policy,[100] *current* purchasing decisions do not accurately account for consumers' "natural" preferences. Governments and large corpora-

[96] The European Central Bank's growing emphasis on decarbonization since 2022 may gradually reduce its investments in the livestock sector.

[97] Schnabel, I. 2021. From market neutrality to market efficiency. Address at the European Central Bank. Available at: https://www.ecb.europa.eu/press/key/date/2021/html/ecb.sp210614~162bd7c253.en.html. ("In the presence of market failures, adhering to the market neutrality principle may reinforce pre-existing inefficiencies that give rise to a suboptimal allocation of resources.")

[98] See notably Ehlermann, C.-D. and L. Laraine, eds. 1997. *Objectives of Competition Law.* Oxford: Hart Publishing; Zimmer, D., ed. 2012. *The Goals of Competition Law.* Cheltenham: Edward Elgar; Vickers, J. 2024. Competition Policy and the Consumer Welfare Standard. *Journal of Antitrust Enforcement*: 1–11.

[99] See notably Barrett Lydgate, E. 2011. Consumer Preferences and the National Treatment Principle: Emerging Environmental Regulations Prompt a New Look at an Old Problem. *World Trade Review* 10(2): 165–188; Iacovides, M. 2014. Marginal Consumers, Marginalized Economics: Whose Tastes and Habits Should the WTO Panels and Appellate Body Consider When Assessing 'Likeness'? *Journal of World Trade* 48(2): 323–349.

[100] Petit, N. and L. Radic. 2024. The Superiority of the Consumer Welfare Standard. EUI Working Paper LAW 2024/20, note 79. Available at: https://papers.ssrn.com/sol3/papers.cfm?abstract_id=5065469

tions influence purchasing decisions. Their policies affect the relative market prices (e.g., through subsidies or transfer pricing) and shape the informational environment that influences consumers' perceptions of the quantity and nature of desirable products (e.g., nutritional policies, advertising and its regulation).

In this vein, the rising consumption of animal-based food products in the Global North does not reflect consumers' natural preferences. Instead, it is a consequence of the declining cost of meat, driven primarily by subsidies for animal-source food production (which are governed by state aid law), the subjection of contract growers to unfair trading practices (which fall notably within the scope of competition law),[101] and public and private efforts to invisibilize animal cruelty on factory farms and discourage consumers from transitioning to alternatives (e.g., nutritional policy, regulation on misleading advertisements, subsidies for the meat and dairy industry's advertising campaigns).[102] In this context, relying on current consumers' meat purchasing habits to shape future policies on the treatment of farmed animals cannot be considered a legitimate expression of "market democracy." Instead, such an approach further entrenches existing market imperfections and underlying power imbalances. As this example shows, economic law does not carry only *distributive* but also *constitutive* consequences, in the sense that it contributes to shaping and orienting economic agents' consumption patterns, preferences, and beliefs.

Last but not least, the argument that it would be *more efficient and rational* to divide the legal spectrum into discrete areas—economic rules and external regulations—each pursuing a single, narrow objective, warrants caution. On the one hand, even the allegedly univocal objectives assigned to economic law (consumer welfare, shareholders' interest, price stability, etc.) are subject to various interpretations, and therefore hard to maximize.[103] On the other hand, the assertion of the superior efficiency of

[101] Leonard, C. 2014. *The Meat Racket: The Secret Takeover of America's Food Business.* New York: Simon & Schuster; see Chap. 5, Ilyana Aït-Ahmed and Solène Kerisit, "Cruelty Under Contract: A Case Study in French Law."

[102] See Chap. 10, Isabelle Doussan, "Using Consumer Law to Improve the Treatment of Farm Animals in Industrial Farming Production" and Chap. 11, Marine Friant-Perrot, "Meat Reduction as an Animal Protection Policy Goal: Exploring the Economic, Health, and Climate Nexus in EU Law."

[103] Spamann, H. and J. Fisher. 2024. Corporate Purpose: Theoretical and Empirical Foundations/ Confusions. In *Corporate Purpose, CSR, and ESG*, Jens-Hinrich Binder, Klaus Hopt, and Thilo Kuntz, eds. 55–81, p. 56. Oxford: Oxford University Press.

siloed policies is generally left unsubstantiated. Commentators often attempt to support this view via the work of the Dutch economist Jan Tinbergen. However, contrary to popular interpretation, his theories do not support the assignment of a single objective to each policy.[104] Instead, he has argued that in complex modern societies, policy instruments inevitably serve multiple objectives, thereby creating space for systematic regulatory discretion.[105] Legal scholars and policymakers are thus faced with a choice: persist in maintaining the fiction of isolated objectives and tools, or acknowledge these interconnections and manage them transparently.[106] In any case, the presumed superior rationality of compartmentalized policy silos remains unproven.

In this context, it is worth considering the possibility of forging a new path for economic law—one that harnesses its transformative potential to address today's most pressing crises, which are deeply embedded in the functioning of the economy and its supporting regulatory frameworks.

3.2 Economic Law as a Cornerstone in the Transition Toward Greater Consideration for Animals

The advancement of animal law, and sustainability law more broadly, in the twenty-first century has been hindered by the longstanding reliance on an outdated model of economic law rooted in the latter half of the twentieth century. Recent efforts to enhance the sustainability and animal

[104] Pottier, A. 2023. Analyse critique de l'usage des arguments économiques en droit de l'environnement. In *Le droit économique de l'environnement: Acteurs et méthodes*, Aude-Solveig Epstein and Marie-Alice Chardeaux, eds. 271–284, p. 283. Paris: Mare & Martin (in French).

[105] Tinbergen, J. 1952. *On the Theory of Economic Policy*. Amsterdam: North Holland Publishing Company.

[106] De Witte, B. 2010. A Competence to Protect: The Pursuit of Non-Market aims through Internal Market Legislation. In *The Judiciary, the Legislature and the EU Internal Market*, ed. Phil Syrpis, p. 44–45, note 32. Cambridge: Cambridge University Press. ("In that context, greater transparency would certainly be welcome. The European legislature could identify the underlying policy goals of internal market legislation more clearly and more systematically within the actual text than is currently the case. If these goals were better articulated, then the interested European citizens could better assess whether, in their opinion, the EU institutions have adequately taken into account the various mainstreaming clauses of the Treaty about environment, fundamental rights, culture, health and social protection and, more generally, whether the legislative act as a whole strikes an adequate balance between the objective of market integration and competing non-market objectives.")

welfare dimensions of sectoral regulations have largely fallen short, as they have failed to confront the fundamental inadequacies of many economic law principles—or, at the very least, of their prevailing interpretations. Unless these foundational rules are reformed, the additional policy layers designed to improve the sustainability of economic transactions are likely to fail.[107] Instead, reformers must address the underlying causes of economic actors' lack of preparedness to embrace sustainability and greater consideration for animal needs and interests. This necessitates a comprehensive re-examination of economic law itself. In essence, law is not merely a tool for addressing the ecological crisis; it has also been a significant driver of the crisis.

As recent research shows, an alternative model of economic law is possible—one that dismantles the artificial separation between economic law and animal law, or more broadly, sustainability law.[108] In this integrated approach, economic law would be actively mobilized to transform the economy, aligning economic development with sustainability, including animal protection. This paradigm shift would incorporate mechanisms to address externalities directly within the design of market regulations and corporate governance. Such an integrated framework would enable the development of more ambitious and effective policies than those that have emerged from the current policy-layering approach to animal law. By addressing the root determinants of economic decisions—rather than merely regulating the downstream consequences of inherently unsustainable business models and technologies—this approach would foster a proactive and comprehensive regulatory environment.

[107] For instance, when analyzing the European Commission's Conclusions of Strategic Dialogue on the Future of EU Agriculture in the EU published in 2023, Baret and Duluins observe that this policy document "proposes corrective measures without addressing the root causes of the problem, as if these were inevitable. (...) This lack of analysis of the causes of dysfunction leads to a policy layering logic, where new proposals are added to existing policies without evaluating or potentially adapting past measures (...). Why are alternative food systems, which are far more sustainable than the dominant model, currently so marginal? The report does not even raise the question. The truth is that they are not sufficiently supported and have even been particularly affected by recent political and economic choices." We translated Baret, P. and O. Duluins. 2024. Dialogue Stratégique: Une Occasion Manquée de Repenser la PAC. *L'Économie Politique* 4(104): 58–68, p. 65–66.

[108] Epstein, A.-S., coord. 2025. La Transformation Écologique du Droit Économique. A report commissioned by the Institut des Etudes et de la Recherche sur la Justice and the French Environmental Protection Agency (Ademe). Available at: https://eco-logic.law/publications-2/la-transformation-ecologique-du-droit-economique/

Additionally, this approach would have the potential to align the scope of animal protection standards with the jurisdiction of covered corporations and the breadth of regulated markets. This would help mitigate a critical weakness of current animal protection statutes: their limited international reach, which encourages race-to-the-bottom policy dynamics.[109] Finally, policies that explicitly and rigorously balance the various interests at stake—anchored in an awareness of the immense costs of ecological inaction and the moral duty to protect sentient beings from suffering—could prove both more ethically sound and economically efficient than the *ad hoc* compromises currently in place as a result of existing power imbalances.[110]

The EU legal landscape is already evolving, with a host of new measures reflecting a better alignment between European economic law and animal protection concerns. For instance, the EU has begun neutralizing certain rules that are incompatible with environmental and animal welfare objectives. For example, the EU has made it clear that agreements between competitors in the agricultural sector are legal insofar as they promote animal health and welfare.[111] The EU is also incentivizing good animal welfare practices in the Common Agricultural Policy[112] and the Common Fisheries Policy.[113] The EU has additionally attempted to integrate animal welfare standards into international trade agreements, by requiring parties to apply and enforce production standards equivalent to the EU's, by

[109] Peters, A. 2021. *Animals in International Law*, p. 532, note 73. Leiden: Brill Nijhoff; Blattner, C. 2019. *Protecting Animals Within and Across Borders: Extraterritorial Jurisdiction and the Challenges of Globalization*. Oxford: Oxford University Press; Sjåfjell, B. and C. M. Bruner. 2019. Corporations and Sustainability. In *The Cambridge Handbook of Corporate Law, Corporate Governance and Sustainability*, B. Sjåfjell and C. M. Bruner, eds. 3–12, p. 6–7. Cambridge: Cambridge University Press.

[110] Even critics of cost-benefit analysis acknowledge that "consideration of the costs and benefits of a program is obviously a requirement for sound policymaking." Source: Rose-Ackerman, S. 2011. Putting Cost-benefit Analysis In Its Place: Rethinking Regulatory Review. *University of Miami Law Review* 6: 335–355, p. 337.

[111] See Article 210a of Regulation (EU) No 1308/2013 of the European Parliament and of the Council of 17 December 2013 establishing a common organization of the markets in agricultural products, 2013 O.J. (L 347) 671–854.

[112] See Chap. 3, Nicolas de Sadeleer and Alice Di Concetto, "The Common Agricultural Policy and Animals: Friends or Foes?"

[113] See Chap. 4, Denis-Sacha Robin, "Protecting Marine Life: Legal Gaps and Policy Contradictions in EU Fishing Regulation."

encouraging the development and adoption of international standards,[114] or, more rarely, by restricting the trade of products that originate from inhumane or unsustainable production methods.[115] Animal welfare concerns have also permeated EU horizontal legislation. For example, animal welfare-related claims are now regulated under rules applicable to misleading advertising,[116] and companies are required to disclose animal welfare policies in their corporate annual reports.[117]

These reforms remain modest, generally encouraging voluntary good practices or, at best, holding economic actors financially accountable for the risks of neglecting sustainability.[118] In the medium term, however, the increasing integration of animal welfare considerations into economic law could pave the way for a reconceptualization of core economic law principles, aligning them more closely with animal protection goals. For example, although the inclusion of animal protection rules in corporate law has

[114] This is the sense of the inclusion of cooperation clauses on animal welfare in free trade agreements—for instance, in the EU–Mexico Trade Agreement and the EU–Chile Interim Trade Agreement.

[115] One notable example is the exclusion of beef products originating from cows raised on feedlots and sheepmeat raised on systems other than pastoral from the scope of the EU–New Zealand Free Trade Agreement, signed in July 2023. Source: Free Trade Agreement between the European Union and New Zealand, O.J. L, 2024/866. See more broadly Nollkaemper, A. 2023. International Law and the Agony of Animals in Industrial Meat Production. *The European Journal of International Law* 34(4): 939–972, p. 945–946 and 957–958; Offor, I. 2020. Animals and the Impact of Trade Law and Policy: A Global Animal Law Question. *Transnational Environmental Law* 9(2): 239–262.

[116] Directive (EU) 2024/825 of the European Parliament and of the Council of 28 February 2024 amending Directives 2005/29/EC and 2011/83/EU as regards empowering consumers for the green transition through better protection against unfair practices and through better information, O.J. (L 2024) 825.

[117] Directive (EU) 2022/2464 of the European Parliament and of the Council of 14 December 2022 amending Regulation (EU) No 537/2014, Directive 2004/109/EC, Directive 2006/43/EC and Directive 2013/34/EU, as regards corporate sustainability reporting, O.J. (L 322) 15–80; Commission Delegated Regulation (EU) 2023/2772 of 31 July 2023 supplementing Directive 2013/34/EU of the European Parliament and of the Council as regards sustainability reporting standards, O.J. (L 2023) 2772.

[118] Recently, legal scholars also insisted on the financial risks taken by business models relying on the cruellest forms of animal exploitation. See, for instance, Parker, C. and L. Sheedy-Reinhard. 2022. Are Banks Responsible for Animal Welfare and Climate Disruption? A Critical Review of Australian Banks' Due Diligence Policies for Agribusiness Lending. *Transnational Environmental Law* 11(3): 603–628.

not yet had a significant impact on corporate governance,[119] recent reforms represent an initial step toward recognizing animals as stakeholders in business operations.[120] As this example illustrates, reformed economic law offers promising opportunities to advance animal welfare—and even animal liberation.

A transformative economic law agenda focused on animal protection also has the potential to open new avenues for research into the legal frameworks that govern human-animal interactions. A notable example is found in recent studies on animal work, which highlight the critical importance of recognizing animals as workers and, consequently, the relevance of labor law in achieving meaningful improvements in their situation.[121] As this example demonstrates, although legal scholarship at the intersection of animal law and economic law is still in its early stages, it nonetheless represents one of the most dynamic and significant frontiers in contemporary legal studies.

Advancing the field of animal law within the broader context of economic law will require a strongly interdisciplinary approach, with the social sciences potentially providing key insights. For example, rather than rejecting cost-benefit analysis entirely, the inclusion of perspectives from animal studies, such as animal ethics and ethology, into this law and economics framework could create promising avenues to advance both animal interests and economic efficiency. At the same time, a solid understanding of economic principles is essential for critically evaluating and challenging the flawed assumptions embedded in purportedly economic arguments that influence the development of animal law. The field of history can also

[119] See notably Ibrahim, D.M. 2007. A Return to Descartes: Property, Profit, and the Corporate Ownership of Animals. *Law and Contemporary Problems* 70(1): 89–115; Smart, J. 2022. Animals as Stakeholders. In *Animals And Business Ethics,* ed. Natalie Thomas, 297–324. London: Palgrave Macmillan; See also Chap. 6, Clémentine Baldon, "Corporate Social Responsibility and Animal Protection."

[120] Agence Symbiocène. 2024. Les animaux, la CSRD et vous. Available at: https://agence-symbiocene.squarespace.com/blog/les-animaux-la-csrd-et-vous

[121] Porcher, J. 2011. *Vivre avec les Animaux: Une Utopie Pour le XXIe Siècle*. Paris: La Découverte; Porcher, J. 2015. Le Travail des Animaux d'Élevage: Un Partenariat Invisible? *Courrier de l'INRA* 65 (in French). Available at: https://hal.science/hal-01222753/file/4-C65Porcher.pdf; Blattner, C.E. 2019. Beyond the Goods/Resources Dichotomy: Animal Labor and Trade Law. *Journal of International Wildlife Law & Policy* 22(2): 63–89; Blattner, C.E., Coulter, K., and W. Kymlicka, eds. 2019. *Animal Labour: A New Frontier of Interspecies Justice?* Oxford: Oxford University Press; Jarrige, F. 2023. *La Ronde des Bêtes: Le moteur animal et la fabrique de la modernité*. Paris: La Découverte (in French).

help reorient markets and corporate structures toward greater consideration for animals, by highlighting the evolving role of animals in the economy[122] and exploring the shifting interpretations of key economic concepts, such as "free trade"[123] or "the corporation."[124]

Even within the legal field, a certain degree of holism is essential for producing robust analyses that can support innovative perspectives on the status of animals in contemporary economies. This calls for moving beyond narrow disciplinary approaches that examine only the rules governing animals within isolated areas of economic law. Instead, scholars must integrate these fragmented insights into a cohesive framework that spans both public and private law. Adopting such a comprehensive perspective is crucial not only for generating theoretical insights into the legal foundations, technical entry points, and obstacles to advancing animal interests within economic law, but also for informing public policy. From a policymaking standpoint, this broader approach is key to designing a balanced policy mix capable of meaningfully embedding animal protection into economic systems and structures.

4 Conclusion

The failure of both welfarist and rights-based approaches to achieve meaningful progress for animals highlights the need for a new legal strategy: one that does not treat animal protection as separate from the economic systems that structure and sustain their exploitation. This chapter develops such a strategy by positioning economic law as a critical yet underexplored frontier for advancing animal interests. Rather than accepting the traditional dichotomy between economic law and so-called external regulations, it argues that EU animal law is fundamentally economic law. It further demonstrates that this economic foundation does not preclude ambitious animal protection policies; instead, the tepid nature of EU animal law stems from regulatory capture rather than economic principles.

A more ambitious animal policy—one that moves beyond limiting egregious harms to animals to embedding animal protection within the

[122] Jarrige, F. 2023. *La Ronde des Bêtes: Le moteur animal et la fabrique de la modernité*, note 122. Paris: La Découverte (in French).

[123] Lang, A. T.F. 2007. Reflecting on "Linkage": Cognitive and Institutional Change in the International Trading System. *Modern Law Review* 70(4): 523–549, note 93.

[124] Ireland P. 2018. Efficiency or Power? The Rise of the Shareholder-oriented Joint Stock Corporation. *Indiana Journal of Global Legal Studies* 25(1): 291–330.

very definition of economic development—requires a thorough reform of economic law. Such reform must begin by rejecting the assumption that non-financial objectives belong solely to so-called external regulations. This contribution shows that integrating animal protection into corporate law and market regulation can result in legal systems that are not only more just but also more economically efficient and legally coherent. Ultimately, it calls on scholars, practitioners, and policymakers to rethink the role of law in shaping an economy where animal interests are not merely acknowledged but meaningfully advanced.

Animals as Economic Inputs

The Common Agricultural Policy and Animals: Friends or Foes?

Nicolas de Sadeleer and Alice Di Concetto

1 Introduction

The EU's Common Agricultural Policy (CAP) came into being in 1962. At this time, the six founding Member States endeavored to restore Europe's agricultural production capacity by establishing the CAP, following World War II and the establishment of the European Economic Community. The CAP was also the EU's first and, for many years, only fully integrated policy. Although the regulations composing the CAP have undergone many reforms since 1962, the CAP's foundational principles have remained virtually unchanged, making it one of the most resilient EU policies. The inclusion of the CAP in the Treaty of Rome, that later

N. de Sadeleer (✉)
UC Louvain, Louvain, Belgium
e-mail: nicolas.desadeleer@uclouvain.be

A. Di Concetto
The European Institute for Animal Law & Policy, Brussels, Belgium
e-mail: contact@animallaweurope.org

A.-S. Epstein, A. Di Concetto (eds.), *Animals in EU Economic Law*, The Palgrave Macmillan Animal Ethics Series,
https://doi.org/10.1007/978-3-032-16171-0_3

became the Treaty of the Functioning of the EU (TFEU), has further provided the legal basis for the EU legislation on farmed animal welfare.[1]

The EU regulates public support to agriculture through a system of subsidies and support programs via the CAP. The CAP is composed of three regulations, which primarily establish rules on distribution of public subsidies—their amount, types, and eligibility criteria for beneficiaries. Through its agricultural subsidies and support programs, the CAP has the power to shape the ways in which agriculture is practiced, thereby determining which types of farming systems are likely to thrive across Europe. The CAP thus influences the diets of 500 million Europeans by incentivizing the supply of certain agricultural outputs.

The EU revises CAP regulations every seven years on average, and until the 2021 reform, CAP expenditures accounted for the majority of the EU's budget. From its outset, the CAP has been successful in achieving food security, so much so that as early as the 1970s, farmers started overproducing food, including beef and butter, famously referred to in the media at that time as "butter mountains." The neoliberal turn of the EU in the 1990s subsequently repurposed the CAP as a driver in increasing the EU balance of trade in agri-food products.

However, the EU failed to address the emergence of negative externalities due to policy orientations in the CAP that have caused the generalization of industrial agricultural production. For instance, while industrial farm animal production has contributed to increasing production volumes in animal-based food, this production method has also caused immense suffering in animals due to routine cruel practices. For instance, the keeping of animals in close confinement is now ubiquitous in the EU, where close to half of egg-laying hens are kept in cages, as well as the majority of breeding sows and calves.[2] Farmed animals also undergo extreme selective breeding practices that undermine their welfare and are additionally subjected to painful mutilations, such as debeaking, dehorning, castration

[1] Consolidated version of the Treaty on the Functioning of the European Union, art. 39, O.J. C 2016/62–63. For more about this topic, see Sowery, K. 2018. Sentient Beings and Tradable Products: the Curious Constitutional Status of Animals Under Union Law. *Common Market Law Review*.

[2] Compassion in World Farming. 2020. End the Cage Age: Why the EU Must Stop Caging Farm Animals. Available at: https://www.ciwf.org/media/7442226/ciwf_cage-age_report_oct-20_english_final.pdf.

without anesthesia, teeth grinding, and tail-docking.[3] Lastly, long distance transports are common in the EU, as well as the use of painful killing methods in industrial slaughterhouses.

The CAP has also caused deleterious effects on the protection of wild animals due to habitat encroachment, and it has caused environmental degradation from intensive agricultural practices.[4] Environmental scientists have found a direct correlation between CAP measures and the eradication of farmland birds.[5] For this reason, the European Environment Agency has identified agriculture as the main driver of habitat destruction in grasslands on the EU's protected natural sites.[6] Conversely, when it supports more extensive agricultural practices, the CAP can play a positive role in the conservation of animals that have thrived as a result of centuries-old agricultural activities that have shaped ecosystems in Europe.[7] Therefore, from an animal protection perspective, animal protection measures in the EU's agricultural policy should aim to do more than just non-intervention in wild ecosystems.

[3] European Commission. 2022. Commission Staff Working Document Fitness Check of the EU Animal Welfare Legislation. SWD (2022) 329 final.

[4] Kmecl, P. and K. Denac. 2018. The effects of forest succession and grazing intensity on bird diversity and the conservation value of a Northern Adriatic karstic landscape. *Biodiversity and Conservation* 27: 2003–2020. https://doi.org/10.1007/s10531-018-1521-2; Zakkak, S., Radovic, A., Nikolov, S. C., Shumka, S., Kakalis, L., and V. Kati. 2015. Assessing the effect of agricultural land abandonment on bird communities in southern-eastern Europe. *Journal of Environmental Management* 164: 171–179. https://doi.org/10.1016/j.jenvman.2015.09.005. The Court of Justice of the EU also found that Natura 2000 areas were affected by agricultural activities. For instance, in the context of the destruction of a bird protected area due to the expansion of corn production, see Case C-96/98, *Commission v France* (1999) ECR I – 8548; and in the context of the degradation of a bird protected area resulting from the illegal deterioration of groundwater, see Case C-559/19, *Commission v Spain* (2021) EU:C:2021:512.

[5] Šumrada, T., Kmecl, P., and E. Erjavec. 2021. Do the EU's Common agricultural policy funds negatively affect the diversity of farmland birds? Evidence from Slovenia. *Agriculture, Ecosystems & Environment* 306. https://doi.org/10.1016/j.agee.2020.107200

[6] European Environment Agency. 2019. The European Environment: State and Outlook 2020. https://www.eea.europa.eu/soer/publications/soer-2020. Accessed 4 December 2024.

[7] De Sadeleer, N. 2023. La Nouvelle PAC: Alliée ou ennemie de la nature sauvage? In *Les vingt ans du règlement sur la législation et la sécurité alimentaire dans l'Union européenne: Bilan et perspectives*, Nathalie De Grove Valdeyron and Marine Friant-Perrot, eds. p. 127. Brussels: Bruylant; De Sadeleer, N., and M. Fauconnier. 2024. L'écoconditionnalité de la politique agricole commune mise à l'épreuve: entre productivisme et dérive environnementale. In *Annuaire de droit de l'Union européen 2024*. Bruxelles: Bruylant.

Over the past three decades, the EU legislature has amended founding EU treaties and CAP regulations in an effort to make the CAP more consistent with citizens' expectations for more sustainable and humane food production systems. First, the EU Member States amended the EU constitutional framework in 2007,[8] by way of the Lisbon Treaty, to give the European Parliament a more prominent role in the decision-making process.[9] To make EU law more consistent with societal aspiration, the EU institutions also mandated the inclusion of environmental[10] and animal welfare objectives in all EU policies, including the EU's agricultural policies[11]—making the CAP the only agricultural policy in the world to take animal welfare into account, at least in theory.

Second, the EU legislature conducted a major reform of the CAP in the early 2000s, to integrate a "multifunctional" approach to the CAP. This approach recognizes agriculture as a provider of non-commodity benefits, also referred to as "public goods," beyond the sole task of producing food.[12] A multifunctional agricultural policy is therefore one that "is capable of [...] conserving nature and making a key contribution to the vitality of rural life, and that responds to consumer concerns and demands as regards food quality and safety, environmental protection, and the safeguarding of animal welfare."[13]

However, the new policy instruments designed to ensure multifunctionality have faced harsh criticism. On one hand, farmers have often lamented the complexity and apparent inconsistency of the growing number of mechanisms. On the other, civil society has questioned the legitimacy and ability of the CAP to counter the spread of industrial production

[8] The EU's founding treaties act as the constitutional treaties of the EU.

[9] As per Consolidated Version of the Treaty on the Functioning of the European Union, art 43(2), 2016 O.J. (C 202) 64–65 [hereinafter TFEU], which provides for the ordinary legislative procedure (TFUE, art 294).

[10] TFUE, art 11, 2016 O.J. (C 202) 53.

[11] TFUE, art 13, 2016 O.J. (C 202) 54.

[12] Daugbjerg, C. and A. Swinbank. 2016. Three Decades of Policy Layering and Politically Sustainable Reform in the European Union's Agricultural Policy. *Governance* 29(2):265–280. https://doi.org/10.1111/gove.12171; Di Concetto, A. 2023. Le bien-être animal dans la Politique Agricole Commune: la prise en compte d'une "attente sociétale." *Revue de droit rural* N°1:14–19. https://www.lexiskiosque.fr/catalog/revue-de-droit-rural/revue-de-droit-rural/n1-2023 (in French).

[13] Conclusions of the Berlin European Council of March 1999, embodying policy on the "Agenda 2000" package cited in European Parliament. 2003. Multifunctional Agriculture and the Reform of the CAP, 22 May 2003.

methods, with advocacy organizations demanding more targeted support and more accountability in the distribution of subsidies.

The latest CAP reform, which took place between 2019 and 2021, partially addressed these issues through a reform of its "delivery model", that is, the way in which funds are distributed to food producers. This reform was the result of lengthy negotiations between the three EU institutions, which took place in 25 trilogues,[14] four super-trilogues, and hundreds of working group meetings. These discussions addressed regulatory and financial rules, aiming to fulfill the numerous objectives of the CAP. A major challenge in the negotiations was aligning the CAP with the objectives presented in the European Green Deal,[15] primarily the achievement of carbon neutrality in the EU by 2050, a goal shared by the Paris Agreement.[16] Given the intertwined nature of the climate and environmental crises, the European Commission adopted, in 2020–2021, a whole host of strategies in a wide array of policy areas to establish bold short- and long-term greenhouse gas climate change objectives, pollution abatement thresholds, and ambitious sustainable development targets.

However, the adoption of the European Green Deal in 2019 followed the publication of proposals for new CAP regulations in 2018. As a result, the objectives of the European Green Deal for the EU's agricultural sector were enacted at a time when negotiations of the CAP reform were well under way. The policy orientations set in the Green Deal and the Farm-to-Fork Strategy[17] were therefore largely excluded from the 2021 CAP

[14] "Trilogues" are negotiations taking place between all three institutions, which are the European Commission, the Council of the EU, and the European Parliament, at the end of the legislative process.

[15] de Sadeleer, N. 2025.The European Green Deal: greenwashing of EU economic policies or a paradigm shift? *European Journal of Risk Regulation*.

[16] European Commission. 2019. Communication from the Commission to the European Parliament, the European Council, the Council, the European Economic and Social Committee and the Committee of the Regions: The European Green Deal. COM/2019/640 final. Available at: https://eur-lex.europa.eu/legal-content/EN/TXT/HTML/?uri=CELEX:52019DC0640

[17] European Commission. 2020. Communication from the Commission to the European Parliament, the Council, the European Economic and Social Committee and the Committee of the Regions: A Farm to Fork Strategy for a fair, healthy and environmentally-friendly food system. COM/2020/381 final. Available at: https://eur-lex.europa.eu/legal-content/EN/TXT/HTML/?uri=CELEX:52020DC0381.

regulations.[18] However, this inconsistency between the Farm-to-Fork Strategy and the 2021 CAP regulations has diminished over time, as the European Commission ended up abandoning most of the measures in the Farm-to-Fork Strategy—starting with the revision of EU farm animal welfare legislation, the adoption of new legislation on food sustainability, and the review of the EU agri-food promotion program. More recently, the European Commission announced the withdrawal of its proposal for a regulation aiming to halve the use of pesticides.[19]

In December 2021, the European Parliament and the Council of the EU eventually adopted three new CAP regulations, which came into force in January 2023. These three regulations—Regulation 2021/2115 Establishing Rules on Support for Strategic Plans to Be Drawn Up by Member States Under the Common Agricultural Policy (CAP Strategic Plans),[20] Regulation 2021/2116 on the Financing, Management and Monitoring of the CAP,[21] and Regulation 2021/2117 Establishing a Common Organisation of the Markets in Agricultural Products[22]—provide the legal framework for the CAP until the next reform, which is expected to take place in 2027 at the earliest.

[18] Although the EU institutions claimed to have included the objectives of the Green Deal, at least to some extent, in the final stages of the legislative process during the trilogues.

[19] European Commission. 2022. Proposal for a regulation of the European Parliament and of the Council on the sustainable use of plant protection products and amending Regulation (EU) 2021/2115. COM/2022/305 final.

[20] Regulation (EU) 2021/2115 of the European Parliament and of the Council of 2 December 2021 establishing rules on support for strategic plans to be drawn up by Member States under the common agricultural policy and financed by the European Agricultural Guarantee Fund (EAGF) and by the European Agricultural Fund for Rural Development (EAFRD) and repealing Regulations (EU) No 1305/2013 and (EU) No 1307/2013 (Regulation on the CAP Strategic Plans), 2021 O.J. (L 435) 1–186.

[21] Regulation (EU) 2021/2116 of the European Parliament and of the Council of 2 December 2021 on the financing, management and monitoring of the common agricultural policy and repealing Regulation (EU) No 1306/2013 (Horizontal Regulation), 2021 O.J. (L 435) 187–261.

[22] Regulation (EU) 2021/2117 of the European Parliament and of the Council of 2 December 2021 amending Regulations (EU) No 1308/2013 establishing a common organisation of the markets in agricultural products, (EU) No 1151/2012 on quality schemes for agricultural products and foodstuffs, (EU) No 251/2014 on the definition, description, presentation, labelling and the protection of geographical indications of aromatised wine products and (EU) No 228/2013 laying down specific measures for agriculture in the outermost regions of the Union, 2021 O.J. (L 435) 262–314.

Because the CAP has played a central role in the history and development of the EU as an economic and political system, political science and economics scholars have produced extensive analyses of this common policy. In comparison, the European legal scholarship has generally neglected the analysis of the legal dynamics at play in the CAP, forsaking how these dynamics more generally reflect the forces at work in the evolution of EU law-making. Yet, the CAP regulation provides, to date, and globally, one of the best ongoing examples of neoliberal legislation.

Similarly, European animal law scholars have paid limited attention to the CAP as a research topic,[23] despite being the main driver of the emergence and proliferation of industrial farm animal production in the EU. As such, CAP regulations are strategic for reformist animal advocates who seek to improve the treatment of wild and domesticated animals exploited for food purposes, but CAP regulations also offer potential incentives to transition away from animal-based production and toward a plant-based economy, which would help secure fundamental rights for animals.

This contribution seeks to address this gap in European scholarship by offering a detailed legal analysis of key animal protection measures within the CAP regulations—focusing mostly on farmed animals but also covering animals living in the wild. Additionally, given that the CAP has evolved over the past six decades into an incredibly complex body of legislation understood by only a few specialists, this chapter also seeks to make the CAP a more accessible subject of research.

This chapter begins by examining how animal protection objectives align with the broader legal and policy goals of the CAP and how these animal welfare objectives have been incorporated, with varying degrees of success, into the CAP regulations. This chapter then presents an overview of wild animal protection measures contained in the CAP regulations. Lastly, this chapter shows that a large array of programs that bolster industrial farm animal production remain in place, undermining efforts to shift away from factory farming to more sustainable food production and consumption models in Europe.

[23] Diane Ryland is an exception. See Ryland, D. 2015. Animal welfare in the reformed Common Agricultural Policy: Wherefore art thou? *Environmental Law Review* 17(1): 22–43.

2 Animal Protection Objectives in the Common Agricultural Policy

Despite historical changes in agriculture and the needs of the EU population, the CAP objectives enshrined in the 1957 Treaty of Rome, which is the founding constitutional treaty of the EU, have remained unchanged. Article 39(1) TFEU lists the general objectives of the CAP regulations as follows:

> to increase agricultural productivity by promoting technical progress and by ensuring the rational development of agricultural production and the optimum utilisation of the factors of production, in particular labour;
>
> thus to ensure a fair standard of living for the agricultural community, in particular by increasing the individual earnings of persons engaged in agriculture;to stabilize markets;
>
> to assure the availability of supplies;
>
> to ensure that supplies reach consumers at reasonable prices.[24]

Overall, the CAP has met its objectives. For instance, compared to 1957, the "availability of supply" is guaranteed in the EU. The subsequent CAP regulations have also ensured food security and "reasonable prices," although the financialization of agricultural commodities, the globalization of food supply chains,[25] and the war in Ukraine have recently undermined such objectives.

The EU legislature and the Court of Justice of the EU have never interpreted such objectives in a way that would update them or expand their scope to include non-socio-economic goals. As a result, objectives such as animal welfare or environmental protection are not considered to be primary objectives of the CAP, even though the EU institutions later integrated these goals into primary and secondary law.

In EU primary law in 1994, Member States first introduced a duty for the EU and the Member States to "pay full regard to the welfare requirements of animals [...]" when "formulating and implementing the Community's agriculture, transport, internal market and research policies" by way of a Protocol on Protection and Welfare of Animals annexed to the Treaty of Amsterdam (a 1997 treaty amending the EU's

[24] Treaty establishing the European Economic Community, 1957, art 39.

[25] More specifically, the EU's dependence on plant-based proteins produced outside the EU has threatened EU food sovereignty, independently from the conflict in Ukraine.

constitutional treaty and expanding the powers of the EU).[26] The wording of the Protocol was later codified in the Treaty of Lisbon to become Article 13 of the TFEU in 2007. However, Article 13, TFEU only sets a general objective that is secondary to the objectives of the CAP established in Article 39, TFEU, as confirmed by the Court of Justice of the EU in several cases.[27]

Another challenging aspect of the inclusion of non-economic goals in the CAP is that the TFEU also requires the Union to take into account many other public interest goals, besides animal welfare, when designing and implementing EU laws. These non-economic, transversal goals include maintaining a "high level" of employment,[28] public health,[29] consumer protection,[30] and environmental protection goals.[31] The multiplicity of these requirements tends to dilute the priority that Member States should grant to each of these respective issues. Additionally, the legal value of these different provisions is unclear due to the use of unspecific language (e.g., "high level of employment," "high level of protection"), which the courts traditionally do not interpret,[32] nor has the legal scholarship analyzed it extensively.

Despite the confusing nature of Article 13 and other provisions that refer to the duty for the EU institutions to take public interest goals into

[26] Protocol on Protection and Welfare of Animals, Treaty of Amsterdam amending the Treaty on European Union, the Treaties establishing the European Communities and certain related acts, 1997 O.J. (C 340) 110.

[27] For instance, C-131/86, *United Kingdom of Great Britain and Northern Ireland v Council of the European Communities*, 1988 ECLI:EU:C:1988:86, para 26; C–189/01, *Jippes and Others*, 2001 ECR I–5718, para 73; T–361/14, *H.B. v Commission,* 2017 ECLI:EU:T:2017:252. See also Sowery, K. 2018. Sentient Beings and Tradable Products: the Curious Constitutional Status of Animals Under Union Law. *Common Market Law Review*; Di Concetto, A. 2022. The Double Edged Sword: International Law and Its Effects on EU Farm Animal Welfare Legislation. *Global Journal of Animal Law, Special Issue: Internal Law and Animal Health and Protection: Persistent Themes, New Prospects for Change* 10(2); Di Concetto, A. 2024. EU Animal Law. In *Nordic Animal Law: Welfare and Rights.* Cambridge: Ethics International Press.

[28] TFEU, art 9, 2016 O.J. (C 202) 53.

[29] Ibid., art 114(3) and 168.

[30] Ibid., art 114(3) and 169.

[31] Ibid., art 11.

[32] For instance, the Court interpreted the duty of Member States to improve living and working standards in the context of social protection as a mere general principle. See Case C-126/86 *Gimenez-Zaera,* 1987 ECR I-3697; and Case C-72/91 *Sloman-Neptune,* 1993 ECR I-887.

account, the EU legislature has gradually integrated animal welfare objectives into secondary law. Since 2003, the EU legislature has listed animal welfare as one of the objectives of the CAP. Specifically, Regulation 2021/2115 elaborated on the objectives set forth in Article 39 TFEU by listing general and specific sub-objectives, including the improvement of animal welfare.[33]

The progressive inclusion of animal welfare as a key objective in the CAP has resulted in the adoption of several rules supporting farm animal welfare, explored below.

3 Farm Animal Welfare Objectives in the CAP

3.1 *Farm Animal Welfare in the "First Pillar" of the CAP*

3.1.1 *The CAP's First Pillar*

The expression "Pillar I" refers to the programs of the CAP funded under the European Agricultural Guarantee Fund (EAGF), which is managed by the European Commission. "Pillar I" measures cover two types of programs.

The first is "Direct Payments," which are subsidies given to producers for the production of agricultural goods (crops and animal products). Since 2003, producers must demonstrate compliance with a series of standards to receive subsidies, including animal welfare standards. Direct payments are calculated based on the number of hectares cultivated and constitute the bulk of the CAP's (and consequently the EU's) budget. Under the 2021 CAP, direct payments remain an essential element in guaranteeing fair income support for farmers,[34] with a total allocated budget of €182.2 billion for the years 2023–2027, representing 72% of the total CAP budget.[35]

[33] Regulation (EU) 2021/2115 of the European Parliament and of the Council of 2 December 2021 establishing rules on support for strategic plans to be drawn up by Member States under the common agricultural policy and financed by the European Agricultural Guarantee Fund (EAGF) and by the European Agricultural Fund for Rural Development (EAFRD) and repealing Regulations (EU) No 1305/2013 and (EU) No 1307/2013 (Regulation on the CAP Strategic Plans), art 6(1)(i), 2021 O.J. (L 435) 28.

[34] Regulation 2021/2115 on the CAP Strategic Plans, Recital 28, 2021 O.J. (L 435) 5.

[35] Fifty-one percent of the total first pillar expenditure is made up of direct payments to farmers. Source: European Commission. Common Agricultural Policy Funds. Available at:

The second CAP program is "Market Measures," which are support programs that aim to facilitate market access to agricultural goods (promotion measures) or relieve the market from agricultural surpluses (market support measures, EU School Schemes).

3.1.2 Animal Welfare Conditionality Rules in the 2021 CAP Regulations

Conditionality rules reduce subsidies for farmers who are found to be non-compliant with a list of environmental, climate, and animal welfare standards listed in Regulation 2021/2115 (CAP Strategic Plans Regulation).[36] In this sense, conditionality rules only facilitate the enforcement of these standards; they do not afford additional payments to farmers.

Compared to previous regulations, the 2021 CAP Strategic Plans Regulation provides stricter environmental standards. The aim of the regulation is to make recipients of direct payments more aware of the need to comply with basic standards, thus making the CAP regulations more consistent with citizens' expectations by "improving consistency of the CAP with the environment, public health, plant health and animal welfare objectives."[37]

There are two types of standards producers must comply with under conditionality rules:

Good Agricultural and Environmental Conditions (GAEC)[38] and Statutory Management Requirements (SMRs).[39]

3.1.2.1 Good Agricultural and Environmental Conditions (GAEC)

GAEC standards aim to maintain "all agricultural areas, including land which is no longer used for production purposes, [...] in good agricultural and environmental condition."[40] Unlike Statutory Management Requirements (SMRs), GAEC can include requirements that are otherwise

https://agriculture.ec.europa.eu/common-agricultural-policy/financing-cap/cap-funds_en#eagf. Accessed 12 February 2025.

[36] Blumann, C (dir.). 2011. *Politique agricole commune et politique commune de la pêche*. Brussels: Éditions de l'Université de Bruxelles.

[37] Regulation 2021/2115 on the CAP Strategic Plans, Recital 42, 2021 O.J. (L 435) 7.

[38] Ibid., Recital 41.

[39] Ibid.

[40] Ibid., art 13(1).

not required under EU law, such as crop rotation. Furthermore, Member States are required to adapt each of the GAEC standards listed in the Regulation,[41] taking into account the specific characteristics of the region concerned, including soil and climatic conditions, existing farming conditions, farming practices, farm size and structure, and land use. Each GAEC standard should also contribute to the achievement of multiple objectives, including animal welfare, even though none of the listed GAEC standards deal directly with animal welfare.

3.1.2.2 Statutory Management Requirements (SMRs)

SMRs standards include three specific farm animal welfare standards, which refer to Articles 3 and 4 of the Directive 2008/119 Laying Down Minimum Standards for the Protection of Calves (SMR 9),[42] Articles 3 and 4 of Directive 2008/120 Laying Down Minimum Standards for the Protection of Pigs (SMR 10),[43] and Article 4 of the Directive Concerning the Protection of Animals Kept for Farming Purposes 98/58 (SMR 11).[44]

Unlike GAECs, conditionality rules on animal welfare do not require farmers to engage in good practices stricter than EU law. Instead, animal welfare SMRs are an enforcement mechanism for existing farm animal welfare legislation.[45] There are also far fewer animal welfare SMRs compared to environmental SMRs, likely because there are more environmental laws in EU law than farm animal welfare provisions.[46]

There exist numerous limitations regarding animal welfare conditionality rules. First, the scope of such SMRs is restricted to three of the five farm animal welfare directives on farm animal welfare.[47] Because poultry

[41] Ibid.

[42] Council Directive 2008/119/EC of 18 December 2008 laying down minimum standards for the protection of calves, 2009 O.J. (L 10) 8.

[43] Ibid., p. 6–7.

[44] Council Directive 98/58/EC of 20 July 1998 concerning the protection of animals kept for farming purposes, 1998 O.J. (L 221) 24.

[45] Di Concetto, A. 2023. Le bien-être animal dans la Politique Agricole Commune: la prise en compte d'une "attente sociétale." *Revue de droit rural* N°1, p. 16. https://www.lexiskiosque.fr/catalog/revue-de-droit-rural/revue-de-droit-rural/n1-2023 (in French).

[46] EU environmental law has undergone spectacular development since the early 1970s.

[47] Apart from Directives 98/58, 2008/119 and 2008/120, the EU also adopted Council Directive 1999/74/EC of 19 July 1999 laying down minimum standards for the protection of laying hens, 1999 O.J. (L 203) 53–57; and Council Directive 2007/43/EC of 28 June 2007 laying down minimum rules for the protection of chickens kept for meat production, 2007 O.J. (L 182) 19–28. Also in that sense, see Ryland, D. 2015. Animal welfare in the

farming typically requires relatively little land, the EU institutions assumed that broiler chicken farms and egg farms should not be included in the scope of the conditional rules because these activities are not eligible for direct payments, which are usually hectare-based. The European Commission also assumed that poultry farms with enough hectares to be eligible for direct payments would likely practice extensive poultry farming and thus would not be covered by Directives 1999/74 on the protection of egg-laying hens,[48] or in Directive 2007/43 on the protection of broiler chickens,[49] which excludes non-industrial farms from their scope. However, the European Commission did not consider that certain Member States might extend the scope of penalties under conditionality rules to payments other than direct payments. For instance, French regulations stipulate that both direct payments under the first pillar and payments under the second pillar may be reduced for lack of compliance with SMRs and GAECs.[50] Excluding the poultry Directives from the scope of the SMRs therefore deprives Member States like France of the ability to penalize poultry producers by reducing CAP subsidies for not complying with farm animal welfare laws.

Second, animal welfare SMRs do not provide high levels of protection for farm animals because they refer to farm animal welfare directives, which are ineffective in protecting animals from cruel but common industry practices. For instance, both the Calves and the Pigs Directives allow extreme confinement through the use of individual pens for calves and cages for sows.[51] The 98/58 Directive Concerning the Protection of Animals Kept for Farming Purposes is also rife with vague wording that produces no regulatory effect on the practical ways in which animals are treated on farms.[52]

reformed Common Agricultural Policy: Wherefore art thou? *Environmental Law Review* 17(1):22–43.

[48] Council Directive 1999/74/EC of 19 July 1999 laying down minimum standards for the protection of laying hens, 1999 O.J. (L 203) 53–57.

[49] Council Directive 2007/43/EC of 28 June 2007 laying down minimum rules for the protection of chickens kept for meat production, 2007 O.J. (L 182) 19–28.

[50] Ministère de l'agriculture et de la souveraineté alimentaire. 2022. La Politique Agricole Commune: "Conditionnalité." https://agriculture.gouv.fr/aides-pac-quest-ce-que-la-conditionnalite. Accessed 26 December 2024 (in French).

[51] Di Concetto, A. 2022. The Double-Edged Sword: International Law and Its Effects on EU Farm Animal Welfare Legislation. *Global Journal of Animal Law* 10(2). https://ojs.abo.fi/ojs/index.php/gjal/article/view/1756.

[52] Ibid.

A third limitation with conditionality rules pertains to the numerous exemptions introduced by the EU legislature. For instance, none of the conditionality rules apply to "small farmers,"[53] which are defined in a way that exempts approximately 40% of farmers in the EU.[54] Overall, conditionality rules only apply to 55% of EU farms.[55]

3.1.2.3 Penalties

A fourth limitation of the conditionality rules has to do with the penalty levels, which are not sufficiently deterrent. Penalties for breaching animal welfare conditionality rules remain low, EU law establishing a 3% reduction in subsidies as a general rule.[56] The likelihood of producers facing penalties for non-compliance with conditionality rules is also low, considering that the Horizontal Regulation only requires Member States to conduct a minimum inspection rate of 1% of farms per year. Furthermore, Member States can and usually do implement an "early warning system," whereby national competent authorities can issue up to three warnings before effectively reducing subsidies.[57] Empirical studies have additionally shown that national competent authorities have been reluctant to inspect farms and impose penalties in the event of fraud.[58] Last but not least, Regulation 2024/1468 later amended the CAP regulations to exempt farms of less than 10 hectares from controls and penalties related to conditionality requirements.[59] As a result, since 2024, these small farms are no longer subject to administrative controls, even though they represent 65%

[53] Regulation (EU) No 1306/2013 of the European Parliament and of the Council of 17 December 2013 on the financing, management and monitoring of the common agricultural policy and repealing Council Regulations (EEC) No 352/78, (EC) No 165/94, (EC) No 2799/98, (EC) No 814/2000, (EC) No 1290/2005 and (EC) No 485/2008, art 92, 2013 O.J. (L 347) 592.

[54] European Court of Auditors. 2018. Special report No 31/2018: Animal welfare in the EU: closing the gap between ambitious goals and practical implementation, p. 12. Available at: https://www.eca.europa.eu/en/publications?did=47557. Accessed 4 December 2024. Also in that sense, Di Concetto, A. 2023. Le bien-être animal dans la Politique Agricole Commune: la prise en compte d'une "attente sociétale." *Revue de droit rural* N°1, p. 17. https://www.lexiskiosque.fr/catalog/revue-de-droit-rural/revue-de-droit-rural/n1-2023 (in French).

[55] Ibid.

[56] Regulation 2021/2116 on the financing, management and monitoring of the common agricultural policy, art 85(1), O.J. (L 435) 247.

[57] Ibid., art 85(3).

[58] Jack, B. 2009. *Agriculture and EU Environmental Law*. London: Routledge.

[59] Article 83(2).

of CAP beneficiaries. In the absence of controls, compliance with GAECs will likely be low. Yet the stakes are as crucial, given that pollution from small farms can cause irreversible damage to the environment.

3.1.3 Schemes for the Climate, the Environment, and Animal Welfare (Eco-Schemes)

Regulation 2021/2115 introduced a new form of payment called "eco-schemes," which compensates farmers for the uptake of good practices.[60] The Regulation further requires that Member States dedicate a minimum of 25% of their budget under the first pillar to the payment of eco-schemes every year.[61] As a result, eco-schemes are a subsidy of a hybrid nature: on the one hand, Member States must offer eco-schemes and must spend a minimum amount of their budget to finance them; on the other, eco-schemes are voluntary for farmers, who can choose not to apply for them.

Unlike direct payments, eco-schemes are afforded annually, meaning that farmers must apply for them every year.[62] Furthermore, Member States can choose which good agricultural practices they wish to support, provided they are "beneficial to the climate, the environment, animal welfare and the fight against antimicrobial resistance"[63] and remunerate practices that go above and beyond those provided in GAEC and SMRs.[64]

To ensure compliance with EU law and harmonization across the 27 Member States, the European Commission approves national eco-schemes lists in accordance with one or more specific objectives, including the "protection of biodiversity, conservation or restoration of habitats or species" and "actions to enhance animal welfare or combat antimicrobial resistance."[65]

Eco-schemes are a welcome development of the 2021 CAP reform in that they incentivize agricultural good practices, such as improved space allowance per individual animal, outdoor access, and increased grazing periods.[66] However, their implementation in national legislation is complex and adds to the regulatory burden of the EU and national competent

[60] Regulation 2021/2115 on the CAP Strategic Plans, art 31, 2021 O.J. (L 435) 40.

[61] Ibid., art 97(1).

[62] Ibid., art 31(7).

[63] Ibid., art 31(3).

[64] Ibid., art 31(5).

[65] Ibid., art 31(4)(e) and (g).

[66] European Commission. 2021. List of potential agricultural practices that the eco-schemes could support under the eco-schemes. Available at: https://agriculture.ec.europa.

authorities, while also making eco-schemes challenging for farmers to navigate. Additionally, because eco-schemes are voluntary measures for farmers, but stricter than conditionality standards, similar to measures remunerated under the second pillar of the CAP, there is a risk of redundancy with already-existing subsidies, both in terms of the substance of measures supported and funding.

3.2 Farmed Animal Welfare in the "Second Pillar" of the CAP

3.2.1 The CAP's Second Pillar

"Pillar II" of the CAP, also called "Rural Development," consists of measures that Member States implement at national level. Unlike direct payments in the first pillar, Pillar II measures provide financial incentives for producers, rather than imposing regulatory requirements. Some of these measures are compulsory (the EU requires Member States to implement them) and some are optional (Member States have the choice to implement them or not). This list of measures constitutes what is known as a national "Rural Development Plan" (RDP), funded through the European Agricultural Fund for Rural Development (EAFRD) and co-financed by the EU and Member States.

In 2021, the EU and its Member States committed to funding the EAFRD with €15.34 billion, making up about 23.2% of the total CAP budget.[67] However, the amount of funding appropriated under the EAFRD could affect the level of funding of Pillar I, given that Regulation 2021/2115 on the CAP Strategic Plans provides that Member States who fund agri-environment measures in excess of 30% of their contribution to the EAFRD are allowed to reduce their funding contribution to eco-schemes.[68]

Member States develop and manage their RDPs, often at regional level, rather than national level, to better account for the specific needs of regions and geographical areas. In this sense, Pillar II fully integrates the

eu/news/commission-publishes-list-potential-eco-schemes-2021-01-14_en#moreinfo. Accessed 1 December 2024.

[67] Rural Development measures benefited from an additional €8070.5 million from the Next Generation EU (NGEU) program to finance economic and social recovery in the wake of the COVID-19 pandemic.

[68] Regulation 2021/2115 on the CAP Strategic Plans, art 97(2), 2021 O.J. (L 435) 82. Member States must, in any case, dedicate at least 25% of the First Pillar to eco-schemes.

subsidiarity principles of EU law, whereby policy "principles are set out at Union level" and "Member States should have ample discretion to lay down specific conditions according to their needs."[69]

3.2.2 Animal Welfare Measures in the Second Pillar

Whereas Pillar I incorporates animal welfare legal standards solely through conditionality, Pillar II includes a category of measures, called "Measure 14," which aims to improve farm animal welfare beyond minimal legal standards, with the objective of developing rural areas. Member States willing to include funding for animal welfare measures can therefore choose to do so. Such measures typically fund programs that improve animal health, e.g., better hoof care or nutrition plans, increase space allowance and outdoor access, or prolong weaning periods.[70] Measure 14 was introduced in the CAP regulations in 2003, at the same time as conditionality requirements on animal welfare.[71] Other measures exist that benefit farm animal welfare in the second pillar, such as Measure 4 for on-farm investment, which can be used to improve outdoor areas for poultry. Measure 1, for the conversion and maintenance of organic agriculture, also benefits animals to the extent that organic standards in animal agriculture are stricter than the legal minimum.[72]

However, Pillar II measures are also subject to significant limitations. First, Measure 14 does not mandate that EU Member States offer animal welfare measures to farmers. As a result, for the 2014–2021 period, only 18 out of the 28 EU Member States[73] had implemented Measure 14 on animal welfare. These 18 Member States allocated €1.5 billion for this

[69] Ibid., Recital 71.

[70] For a detailed analysis of such measures, see RSPCA. 2018. Into the Fold Targeted Financial Support to Improve Farm Animal Welfare.

[71] Council Regulation 1257/1999 on support for rural development from the European Agricultural Guidance and Guarantee Fund (EAGGF) and amending and repealing certain Regulations, 1999 O.J. (L 160) 80–102.

[72] Duval, E., von Keyserlingk, MAG., and B. Lecorps. 2020. Organic Dairy Cattle: Do European Union Regulations Promote Animal Welfare? *Animals* 10 (10):1786; Duval, E. 2022. Le droit de l'Union relatif à l'agriculture biologique: la promotion d'un meilleur bien-être des animaux d'élevage? In *Bien-être et normes environnementales*, ed. Isabelle Michallet. Paris: Mare & Martin (in French); Di Concetto, A. 2023. Farm Animal Welfare and Food Information for EU Consumers: Harmonizing the Regulatory Framework for More Policy Coherence. *European Journal of Risk Regulation* 1/2023.

[73] Accounting for a total of 70 Rural Development Programs out of the 118 programs in the EU. Source: European Network for Rural Development. RDP Summaries. https://

measure, which represents only 1.5% of the total expenditure of Pillar II, an amount much lower than the programmed spending, indicating that Member States have underspent on animal welfare measures.

A second limitation of Measure 14 is that it is not tied to strict criteria with effective reporting and inspection mechanisms. For instance, a 2018 report by the European Court of Auditors found that a pig producer was receiving payments under Pillar II's Measure 14 despite violating the ban on the routine tail-docking of pigs.[74]

A third limitation is that animal welfare payments often diverge from their intended target through the modernization of intensive livestock production systems that, for instance, may only marginally improve animal welfare.

Lastly, as with other CAP funding,[75] there exist instances of fraud due to the EU's limited enforcement powers. For instance, on 24 October 2024, the European Public Prosecutor's Office (EPPO) in Croatia arrested nine individuals suspected of defrauding the EU for more than €9 million in CAP funds that were intended for improving animal welfare in the pig sector.[76]

4 The Protection of Habitats and Wild Animals in the CAP

4.1 *Environmental Protection Objectives in the CAP*

The CAP regulations have integrated a wealth of environmental protection and biodiversity conservation measures over the past six decades,

ec.europa.eu/enrd/policy-in-action/rural-development-policy-in-figures/rdp-summaries-2014-2020_en.html. Accessed 1 December 2024.

[74] European Court of Auditors. 2018. Special report No 31/2018: Animal welfare in the EU: closing the gap between ambitious goals and practical implementation, p. 44. https://www.eca.europa.eu/en/publications?did=47557. Accessed 4 December 2024.

[75] Gebrekidan, S., Apuzzo, M., and B. Novak. 2019. The Money Farmers: How Oligarchs and Populists Milk the EU for Millions. *The New York Times*, 3 November. https://www.nytimes.com/2019/11/03/world/europe/eu-farm-subsidy-hungary.html. Accessed 1 December 2024.

[76] European Public Prosecutor's Office. 2024. Croatia: Nine Arrested in Probe into Multi-Million Subsidy Fraud in Pig Farming. https://www.eppo.europa.eu/en/media/news/croatia-nine-arrested-probe-multi-million-subsidy-fraud-pig-farming. Accessed 1 December 2024.

especially since 2003 and the advent of a multifunctional approach to the EU's agricultural policy.

Green payments first emerged as a form of revenue support for farmers to compensate for budgetary cuts to direct payments introduced in the 1992 CAP reform.[77] In 1999, the EU institutions simplified these green payments and reintroduced incentives for more extensive forms of farming in Pillar II. Since 1999, the CAP thus includes agri-environmental measures, in addition to national rural development funds which are also intended to favor the environment.

The 2021 CAP reform did not bring major change to the structure of environmental subsidies—the so-called green payments. The subsequent section provides an overview of the key measures of the CAP specifically aimed at protecting habitats and wild animals, ensuring that this section is not a repetition of existing literature.[78]

4.2 *Wild Animal and Habitat Protection in the First Pillar of the CAP*

4.2.1 *Conditionality Rules*

Under the First Pillar, four main conditionality rules address the protection of habitats and wild animals: two Good Agricultural and Environmental Conditions (GAEC), and two Statutory Management Requirements (SMRs).

Regulation 2021/2115 on the CAP Strategic Plans requires producers to comply with a GAEC standard (GAEC 8) whose main objective is the "maintenance of non-productive features and area to improve on-farm biodiversity."[79] More specifically, this standard requires that producers maintain a minimum of:

[77] De Sadeleer, N. and C.-H. Born. 2004. *Droit international et communautaire de la biodiversité*. Paris: Dalloz.

[78] For instance, see De Sadeleer, N. 2023. La Nouvelle PAC: Alliée ou ennemie de la nature sauvage? In *Les vingt ans du règlement sur la législation et la sécurité alimentaire dans l'Union européenne: Bilan et perspectives*, Nathalie De Grove Valdeyron and Marine Friant-Perrot, eds. 97–138. Brussels: Bruylant.

[79] Regulation 2021/2115 on the CAP Strategic Plans, Annex III, 2021 O.J. (L 435) 150.

either 4% of non-productive areas (honey set-aside) and/or biodiversity elements (hedges, copses, trees, ponds, etc.) in relation to their arable land area;
or 3% of non-productive areas combined with 4% of catch crops and/or nitrogen-fixing crops (peas, beans, clover, lupins, alfalfa, vetches, etc.), without plant protection treatment.

Member States were expected to comply with GAEC 8 as early as 2023. However, the European Commission granted an exceptional derogation, allowing Member States not to comply with GAEC 8 in 2023. This derogation was further extended to 2024 and, in May of that year, the European Parliament and the Council of the EU eventually decided to codify a significantly watered down version of GAEC 8.[80] The codified version of GAEC 8 only requires farmers to maintain topographical features, prohibits farmers from trimming hedges and trees during the bird nesting and breeding season, and provides optional measures to prevent invasive plant species.[81] The final version of GAEC 8 is thus woefully inadequate to protect nature. As a result of this watering down of the legislation, farmers are now authorized to make unrestricted use of their arable land available for food production and will no longer be required to devote a minimum proportion of their arable land to non-productive areas, such as set-aside land, hedges, and low walls. The changes brought to GAEC 9 are therefore likely to accelerate the sharp decline in biodiversity in rural areas.

Another GAEC requires producers to comply with a "Ban on converting or ploughing permanent grassland designated as environmentally-sensitive permanent grasslands in Natura 2000 sites," which are protected natural areas under EU law (GAEC 9).

In addition to these two GAEC standards, Regulation 2021/2115 on the CAP Strategic Plans includes two SMRs relevant to wild animals: SMRs 3 and 4. These two SMRs stipulate penalties in the form of subsidy reduction for producers who do not comply with Directive 2009/147/EC on the Conservation of Wild Birds (Birds Directive)[82] and Directive

[80] Regulation (EU) 2024/1468 of the European Parliament and of the Council of 14 May 2024 amending Regulations (EU) 2021/2115 and (EU) 2021/2116 as regards good agricultural and environmental condition standards, schemes for climate, environment and animal welfare, amendment of the CAP Strategic Plans, review of the CAP Strategic Plans and exemptions from controls and penalties.

[81] Ibid., Annex III.

[82] Ibid., Annex III.

92/43/EEC on the Conservation of Natural Habitats and of Wild Flora and Fauna (Habitats Directive).[83] However, these two SMRs suffer from the same shortcomings as the SMRs related to animal welfare: the language used in the Directives is not specific enough and enforcement gaps are too numerous to ensure adequate levels of protection to animals.[84]

4.2.2 *Wild Animal Protection in the Eco-Schemes*

The European Commission's published list of examples of agricultural practices eligible for support via eco-schemes is relatively expansive and includes a few examples of practices that are likely to benefit wild animals. Relevant examples include organic farming practices, the use of "mixed species/diverse sward of permanent grassland for biodiversity purpose (pollination, birds, game feedstocks)," and the "establishment and maintenance of high-biodiversity silvo-pastoral systems."[85]

4.3 *Wild Animal and Habitat Protection in the Second Pillar of the CAP: Environmental, Climate-Related, and Other Management Commitments (ECMC)*

One category of CAP measures that supports wild animal protection is classified as "Environmental, Climate-Related, and Other Management Commitments" (ECMC). These voluntary measures are intended to compensate farmers for income losses and costs associated with implementation. Farmers may apply to receive these funds for a period of five to seven years.[86]

Like eco-schemes, but unlike GAEC standards, ECMC remunerates voluntary practices that go above and beyond common industry practices. As a result, only practices that go beyond the SMRs and GAECs standards and the relevant minimum requirements relating to the use of fertilizers

[83] Ibid., Annex III.

[84] De Sadeleer, N. 2023. La Nouvelle PAC: Alliée ou ennemie de la nature sauvage? In *Les vingt ans du règlement sur la législation et la sécurité alimentaire dans l'Union européenne: Bilan et perspectives*, Nathalie De Grove Valdeyron and Marine Friant-Perrot, eds. p. 117. Brussels: Bruylant.

[85] European Commission. 2021. List of potential agricultural practices that the eco-schemes could support under the eco-schemes. https://agriculture.ec.europa.eu/news/commission-publishes-list-potential-eco-schemes-2021-01-14_en#moreinfo. Accessed 1 December 2024.

[86] Regulation 2021/2115 on the CAP Strategic Plans, art 70(6), 2021 O.J. (L 435) 65.

and plant protection products or animal welfare are eligible for funding as ECMC.[87] Furthermore, as per the subsidiarity principle that governs the CAP, the content of ECMC measures is defined by Member States, either at national or regional level—provided that these measures align with the general objectives of Regulation 2021/2115 on the CAP Strategic Plans. Consequently, national competent authorities must ensure that eco-schemes and the ECMC do not overlap but instead complement each other effectively.

5 Measures Undermining Animal Protection

Animal-source producers receive very few direct subsidies from the EU. Rather, the meat and dairy industry receives support through a large array of indirect subsidies.[88] These support programs disproportionately benefit industrial animal agriculture compared to the animal protection measures described above.

Several measures under Pillar II undermine efforts to improve animal welfare. This is particularly true of certain subsidies for on-farm investments (Measure 4 mentioned above), which may end up funding the construction of industrial infrastructure, including warehouses for intensive chicken farming, bio-gas tanks, or even manure lagoons to help producers comply with environmental regulations. These subsidies have the effect of incentivizing industrial farm animal production by externalizing the significant production costs necessary to maintain factory farming operations.

The CAP also provides funding for the EU's agricultural promotion policy, which aims to promote agri-food products made in the EU. To achieve this goal, the European Commission's European Research Executive Agency (EREA) grants funds to inter-trade organizations so they can run advertising campaigns inside and outside the EU.[89] These campaigns primarily advertise products, not brands—when they do advertise brands, advertising materials must display multiple brands. Currently, all EU-based inter-trade organizations of all agri-food sectors are eligible

[87] Ibid., art 70(3) a) and b).

[88] Di Concetto, A. 2023. Le bien-être animal dans la Politique Agricole Commune: la prise en compte d'une "attente sociétale." *Revue de droit rural* N°1, p.16. https://www.lexiskiosque.fr/catalog/revue-de-droit-rural/revue-de-droit-rural/n1-2023 (in French).

[89] European Commission, European Research Executive Agency. Promotion of Agricultural Products. https://rea.ec.europa.eu/funding-and-grants/promotion-agricultural-products-0_en. Accessed 26 December 2024.

to apply for these promotional grants, regardless of the type of products, or their method of production.[90] In the 2016–2020 period, the EU granted €252.4 million to inter-trade organizations so that they could promote meat and dairy products.[91]

Overall, the amount dedicated to promoting animal-source food is disproportionately high compared to the amount dedicated to promoting fruits and vegetables.[92] In the same funding period, 32% of the total spending of €776.7 million went to promote meat and dairy products, as opposed to 19% used for fruits and vegetables.[93] Meat and milk further benefited from "mixed" campaigns, which promoted several products, including animal-source food. The cost of these mixed campaigns is estimated at €215 million (28% of the EREA funds).[94]

Over the previous financial period, EU-funded campaigns[95] resulted in encouraging the production and consumption of inhumane food products. For instance, the "Let's Talk About Pork" campaign, which has been running in France, Spain, and Portugal, encourages young consumers to eat more pork. The campaign touted pork production in these countries as an animal-friendly industry on the grounds that industrial hog farms are compliant with high animal welfare standards, without mentioning that EU and national animal welfare legislation does in fact allow inhumane practices, including mutilations and extreme confinement. As a result of these low standards, more than 90% of pigs in France and Spain are raised on farms where sows are kept in cages, piglets undergo castration without

[90] All products listed in Annex I of the TFUE, including all animal source food, are eligible for CHAFEA funds, as per Article 5 of Regulation 144/2014 of the European Parliament and of the Council of 22 October 2014 on information provision and promotion measures concerning agricultural products implemented in the internal market and in third countries and repealing Council Regulation (EC) No 3/2008, 2014 O.J. (L 317) 56–70.

[91] Greenpeace. 2021. Marketing Meat: How EU Promotional Funds Favor Meat and Dairy. Available at: https://www.greenpeace.org/static/planet4-eu-unit-stateless/2021/04/20210408-Greenpeace-report-Marketing-Meat.pdf.

[92] Also see Chap. 11, Marine Friant-Perrot, "Meat Reduction as an Animal Protection Policy Goal: Exploring the Economic, Health, and Climate Nexus in EU Law."

[93] Greenpeace. 2021. Marketing Meat: How EU Promotional Funds Favor Meat and Dairy. Available at: https://www.greenpeace.org/static/planet4-eu-unit-stateless/2021/04/20210408-Greenpeace-report-Marketing-Meat.pdf.

[94] Ibid.

[95] These funds were then managed by the EU's Consumers, Health, Agriculture and Food Executive Agency (CHAFEA).

anesthesia, as well as teeth-clipping and tail docking, and all throughout the production chain these animals are deprived of outdoor access.[96]

In its Farm to Fork Strategy, the European Commission had committed to conduct a review of the EU agri-food promotion program "to support the most sustainable, carbon-efficient methods of livestock production."[97] The Commission ultimately never conducted this review and spending levels for meat and dairy products in 2022 and 2023 remained the same as in previous years.[98] In 2022, promotion campaigns for animal-source food products represented more than 50% of the promotion programs approved for that year,[99] and 25.3% of the approved applications were for meat and dairy products. In 2023, campaigns concerning animal-source products represented 57% of the promotion programs, including 36.8% exclusively covering meat and dairy products. Overall, between 2022 and 2023, the budget allocated to the promotion of animal-source products increased by 37%, while the budget allocated to promoting fruit and vegetables fell by 17%.[100]

5.1 EU School Milk Scheme

The CAP funds another animal-source food program, the EU School Fruit, Vegetables, and Milk Scheme (referred to hereafter as the "School Milk Scheme.") This program was launched in 2017, just two years after milk quotas had ended. The School Milk Scheme is intended to "support

[96] Ninetey-nine percent, 98.5%, and 65% of piglets undergo tail docking respectively in France, Spain, and Portugal. Source: European Commission (DG SANTE). Audits on Measures to Prevent Tail Biting and Avoid Tail-Docking of Pigs in Spain (2017), France (2019), and Portugal (2019).

[97] European Commission. 2020. Communication from the Commission to the European Parliament, the Council, the European Economic and Social Committee and the Committee of the Regions A Farm-to-Fork Strategy for a Fair, Healthy and Environmentally-Friendly Food System. COM (2020) 381 final, p. 7–8.

[98] European Commission. 2023. Commission implementing decision of 14 November 2023 on the financing of information and promotion actions concerning agricultural products on the internal market and in third countries and the adoption of the work program for 2024. C(2023) 7602 final.

[99] Meat and dairy products, and mixed produced basket.

[100] The authors thank Anatole Poinsot for conducting this detailed analysis, which covers Simple and Multi Programs on the internal market and third countries, but excludes the applications co-financed by the EU and the applicants. This analysis uses data for Simple Programs extracted from the European Commission's website and the list of EU-funded Multi-Programs provided on request by the European Commission.

the distribution of products, educational measures and information measures."[101] In economic terms, support afforded to milk in the School Milk Scheme provides a privileged market for milk producers through the subsidization of public procurement of milk and dairy products for school cafeterias across the 27 Member States. Additionally, because the milk measures in the School Milk Scheme target young children, mostly in primary schools,[102] this program further enables producers to influence food choices and lifelong consumption habits of upcoming generations of European citizens.

Over the 2023–2029 period, the annual budget allocated to the School Milk Scheme is €90.1 million per school year. This number only represents the cost of the subsidies that directly benefit the sector and does not take into account the long-term health effects of consumer food choices of the 15 to 19 million children who are offered a total of around 130 million liters of milk every year.[103]

Additionally, only a few Member States give priority to the environment when implementing the School Milk Scheme, even if a majority of them privilege organic milk.[104] However, standards in organic milk production are too lax to ensure high levels of animal protection. Specifically, the Organic Regulation does not establish a cap on the number of animals per holding, and thus hardly limits the concentration and specialization in the dairy sector, two features that are associated with intensive and environmentally deleterious practices,[105] as well as lower levels of cow welfare.

[101] European Commission. School Scheme Explained. https://agriculture.ec.europa.eu/common-agricultural-policy/market-measures/school-fruit-vegetables-and-milk-scheme/school-scheme-explained_en. Accessed 26 December 2024.

[102] European Commission. EU School, Fruit, Vegetable, and Milk Scheme: Monitoring Data. Available at: https://agridata.ec.europa.eu/extensions/SchoolSchemeMonitoring/SchoolSchemeMonitoring.html. Accessed 26 December 2024.

[103] European Commission. 2021. Final Minutes—Civil Dialogue Group Milk, 26 February 2021. Presentation by Maria Giulia Medico. Brussels: Directorate-General for Agriculture and Rural Development. Available at: https://agriculture.ec.europa.eu/system/files/2021-04/cdg-milk-2021-02-26-minutes_en_0.pdf.

[104] Ibid.

[105] CEAS Consultants (Wye) Ltd. and European Forum on Nature Conservation and Pastoralism. 2000. The Environmental Impact of Dairy Production in the EU: Practical Options for the Improvement of the Environmental Impact. Final Report for the European Commission.

6 Conclusion

The new policy instruments designed to ensure the "multifunctionality" of the CAP—the idea that the CAP should do more than just incentivizing high volumes of production—have faced harsh criticism over the past two decades. On the one hand, farmers deplore the complexity of ever more numerous, seemingly inconsistent mechanisms. On the other, civil society has questioned the role of the CAP in spreading industrial production methods, with advocates demanding better targeted support and more accountability in the distribution of subsidies. Despite such growing discontent, successive reforms of the CAP have failed to reduce the ecological footprint of agriculture in the EU or achieve significant improvements in animal welfare.[106]

Consistent with past reforms, the 2021 CAP reform primarily sought to remedy these numerous imperfections. However, the dilution of the GAEC standards in 2024 and the low ambition of National Strategic Plans have undermined this objective. Another limitation of the 2021 CAP reform is that it mainly consisted of a reform of the delivery model of the CAP—the way in which the money should be distributed to farmers—rather than through a root and branch reassessment of the foundations of the policy itself. This new delivery model provides more independence to Member States, in virtue of a reinforced subsidiarity principle, by allowing Member States to list the specific measures that they want to implement on their territory. This considerable latitude left to Member States is likely to create discrepancies in the ways in which agriculture is subsidized from one Member State to the other. As a result, the CAP remains incredibly complex and ineffective, and may additionally cause distortions in competition across the EU market. The eco-schemes are an example of a CAP measure that fails in these multiple and overlapping key areas.

[106] European Court of Auditors. 2018. Special report No 31/2018: Animal welfare in the EU: closing the gap between ambitious goals and practical implementation; European Court of Auditors. 2020. Special Report 13/2020: Biodiversity on farmland: CAP contribution has not halted the decline; European Court of Auditors. 2021. Special Report 16/2021: Common Agricultural Policy and climate: Half of EU climate spending but farm emissions are not decreasing; European Court of Auditors. 2021. Special Report 20/2021: Sustainable water use in agriculture: CAP funds more likely to promote greater rather than more efficient water use; European Court of Auditors. 2021. Special Report 21/2021: EU funding for biodiversity and climate change in EU forests: positive but limited results.

On the specific issue of animal protection, while environmental and animal welfare concerns have made their way into primary law and secondary law, including the CAP regulations, the EU legislature has failed to specify the concept of animal welfare in the context of the CAP. As a result, regulatory standards related to animals are few and minimal. Because of inter-state competition, Member States typically do not attempt to go beyond the minimum standards set in EU law. Lastly, CAP regulations do not provide rules on how funds should be divided between animal welfare measures and environmental measures, and consequently, animal welfare measures tend to receive very limited budgetary support.

It is therefore unlikely that the new CAP regulations will properly facilitate a transition in the EU animal agriculture industry—from industrial farm animal production and toward more humane, sustainable, and secure food systems.[107]

[107] In that sense, see European Court of Auditors. 2024. Special report 20/2024: Common Agricultural Policy Plans.

Protecting Marine Life: Legal Gaps and Policy Contradictions in EU Fishing Regulations

Denys-Sacha Robin

1 Introduction

The term "industrial fishing" is widely used by fisheries operators, the media, nonprofits, and policymakers, in a variety of related contexts. However, the various practices that comprise industrial fishing are not well defined in law, whether EU law or French law. While French law does provide administrative classifications for different types of commercial fishing activities, such classifications do not specifically define "industrial fishing." Instead, French law defines "large-scale fishing" as fishing practices that require the use of vessels with either a gross tonnage equal to or greater than 1200 UMS (Universal Measurement System), or at least 200

D.-S. Robin (✉)
Paris Nanterre University, Nanterre, France
e-mail: dsrobin@parisnanterre.fr

A.-S. Epstein, A. Di Concetto (eds.), *Animals in EU Economic Law*, The Palgrave Macmillan Animal Ethics Series,
https://doi.org/10.1007/978-3-032-16171-0_4

UMS if the vessels are typically away from their home ports or supply points for more than 20 days.[1]

Despite a lack of regulatory definition, industrial fishing is generally understood by policymakers as the activity of fishing vessels ranging from 16 to 140 meters in length, and equipped with modern fishing gear, enabling them to fish in very large quantities and to keep their catch on board for several days or weeks.[2] The expression "industrial fishing" has been coined to describe practices that have developed to its present scale during the 1990s, which is relatively recent in light of the millennia-long history of fishing. The term "industrial fishing" thus came into usage to underscore more modern, industrial fishing methods, as opposed to what is known as "artisanal" or "traditional" fishing. Unlike industrial fishing, traditional fishing is typically carried out by a higher number of vessels of smaller size (from five to 16 meters). Additionally, vessels used for traditional fishing remain for a maximum of 96 hours within a 12-mile zone (covering the geographical scope of territorial sea for "inshore" fishing purposes) or in a zone between 12 and 200 miles from the coast (covering the geographical scope of the exclusive economic zone for the purpose of "offshore" fishing).[3]

Beyond the ecological footprint of industrial fishing, this practice primarily appears to be a specific phenomenon combining at least three of the following characteristics. First, fishing vessels have significant autonomy, allowing them to cover distances of several hundred nautical miles over the span of weeks. According to a study published in the journal *Science* in 2018, whose authors collected tracking data from around 70,000 fishing vessels between 2012 and 2016, vessels were estimated to have covered

[1] See article 12 of the Order of 11 August 2020 on navigation genres, issued in application of Decree no. 2015–723 of 24 June 2015 on the issue of maritime vocational training qualifications and the conditions for performing duties on board vessels armed for trade, pleasure, fishing, and marine cultivation. The "UMS" unit was adopted by the International Maritime Organization to replace the "tonneau" with 200 UMS corresponding to 150 tonneaux.

[2] See, for example, the definitions proposed by the Institut Français de Recherche pour l'Exploitation de la Mer (Ifremer) for classifying different types of fishing. Available at: https://peche.ifremer.fr/Le-monde-de-la-peche/La-peche/Laquelle. Accessed 20 August 2023.

[3] Ibid. Of the 71,631 fishing vessels registered in the EU, around 70% are vessels of 15 meters or less, making up the so-called artisanal fleet. See "European fishing in figures" (situation as at April 2023) on the European Parliament website. Available at: https://www.europarl.europa.eu/factsheets/fr/sheet/122/la-peche-europeenne-en-chiffres. Accessed 20 August 2023.

460 million kilometers (equivalent to 73% of the oceans' surface area) with an energy consumption of 19 billion kWh.[4] The second major characteristic of industrial fishing is the enormous quantity of fish caught by the vessels, which can reach up to 600 tons in a single day for the biggest "factory ships" (compared with a few hundred kilograms to two tons for a small-scale fishing vessel). Operators also typically preserve their catch on board, by freezing it, or processing it into meal or surimi paste, for example. Lastly, industrial fishing activities are defined by the methods and gear used by the vessels, which systematically employ large-scale "dragging gear."

Industrial fisheries operators typically use seabed-scraping gear such as dredges, bottom trawls, and Danish seines, which are weighted nets or traps, towed by one or two vessels. This type of equipment scrapes the seabed, often up to 10 centimeters deep, over areas of several square kilometers. Operators also utilize equipment such as pelagic trawls and seines, which are nets several hundred square meters in size that do not scrape the seabed, but typically concentrate very large quantities of fish and other marine animals, without distinction as to a target or non-target species.[5] In addition to these large-scale capture methods, operators use many technologically advanced tools: GPS devices, probes, and sonar or acoustic cameras are said to have doubled the fishing capacity of equipped vessels, which, according to some studies, are therefore likely to catch as many fish as in years past, even though stocks have declined, generating an observation bias as to their availability and renewal.[6] This enhanced capability of modern industrial fishing vessels also explains why the downsizing of national fishing fleets, in terms of numbers of vessels, does not necessarily lead to a reduction in catching capacity, which therefore continues the diminishment of already depleted fishing stocks.

[4] Kroodsma. D.A., Mayorga, J., Hochberg, T., Miller, N.A., Boerder, K., Ferretti, F., Wilson, A., Bergman, B., White, T.D., Block, B.A., Woods, P., Sullivan, B., Costello, C., and B. Worm. 2018. Tracking the Global Footprint of Fisheries. *Science* 359: 904–908.

[5] See, for example, the description of fishing gear on the Ifremer website. Available at: https://peche.ifremer.fr/Le-monde-de-la-peche/La-peche/comment/Les-engins. Accessed 20 August 2023.

[6] These are the conclusions of several researchers who are members of the Sea Around Us initiative from The University of British Columbia. Lee, L. 2019. New Technology Allows Fleets to Double Fishing Capacity - and Deplete Fish Stocks Faster. https://news.ubc.ca/2019/09/16/new-technology-allows-fleets-to-double-fishing-capacity-and-deplete-fish-stocks-faster/. Accessed 20 August 2023.

Over the past three years, industrial fishing has received increased media and political attention.[7] The public has gradually become more familiar with the issue, thanks in large part to the actions of several key environmental protection organizations, along with the concerns of fishers themselves, and reports published by various public authorities. As more evidence emerges, the consensus becomes clearer: the direct and indirect impacts of industrial fishing have been and continue to be catastrophic for marine environments and all life that depends on the oceans, including humans and other terrestrial creatures.

In terms of the direct impacts of industrial fishing on marine ecosystems, at least four salient consequences can be identified, with many others becoming apparent the closer one looks. The first impact concerns overfishing and the problem of fish stock renewal. According to a 2022 report by the United Nations' Food and Agriculture Organization (FAO), some 90 million tons of fish are caught worldwide every year (including around 3.5 million in the European Union), and 90% of global fish stocks are either fully exploited or overexploited.[8] The situation is similar in the EU, where the European Court of Auditors highlighted in 2020 that despite some progress and a reduction in overfishing in the Atlantic, a loss of biodiversity and even overexploitation has been occurring in European waters, particularly in the Mediterranean.[9] These issues are further

[7] See, for example, Torgemen, É. 2019. Ces chalutiers géants qui vident la Manche et inquiètent les petits pêcheurs. Le Parisien, 20 December; Moullot, P. 2022. Union européenne: la senne démersale, technique de pêche industrielle que veulent bannir ONG et professionnels. Libération, 29 September; Kempinski, F. B. 2022. Aires marines protégées: la guerre des pêches. Les Echos, 17 October; Romano, V. 2023. EU Biodiversity Plan Aims to Ban Bottom Trawling in Protected Areas. Euractiv, 30 January.

[8] FAO. 2022. The State of World Fisheries and Aquaculture 2022. Towards a blue transformation, p. 3 and 50. Available at: https://doi.org/10.4060/cc0461en. See also, for example, the fact sheet "Will there still be fish in the sea in 2050?" published by Ifremer in February 2021, which describes scientific stock assessment methods and their limitations. Available at: https://www.ifremer.fr/sites/default/files/2022-10/fiche_poissons_surpeche.pdf. Accessed 20 August 2023.

[9] European Court of Auditors. 2020. Marine environment: EU protection is wide but not deep. Special Report 26/2020, p. 36–47. See also, European Environment Agency. 2020. Marine Messages II: Navigating the Course Towards Clean, Healthy and Productive Seas Through Implementation of an Ecosystem-Based Approach. EEA Report No. 17/2019, p. 82.

compounded by the persistence, even within the EU, of illegal, unreported, and unregulated fishing or deliberate overshooting of certain quotas.[10]

The second direct impact of industrial fishing on marine ecosystems concerns unwanted catches, either of non-targeted fish species (or immature fish, for example), or of protected marine animals. According to various estimates,[11] these unwanted catches represent around 10% of the annual catch. In practice, for example, 30 to 40% of a trawl net's catch may be non-target species. Some scientists estimate that at least 90 endangered species are caught in trawl nets worldwide.[12] The third direct environmental impact stems from the massive use of towed gear, which results in the destruction of the seabed. The removal of sediments and the crushing or burial of species or their habitat leads to a radical change in the benthic ecosystem (the ecological region at the lowest level of the sea), including chemical modification and acidification. According to scientific studies, 66% of the shallow seabeds (depth between 0 and 200 meters) in the Bay of Biscay and along the Iberian coasts, and 62% of the shallow bottom of the Channel and the North Sea, were subjected to trawling pressure between 2013 and 2018. This use of towed gear has thus consequently led to a degradation of 77 to 88% of the benthic ecosystem in the English Channel, and around 90% of the seabed in the areas studied in the Mediterranean.[13]

[10] See, for example, the case of several European companies affected by fines. Kerbastard, G. and T. Peigné. 2021. Pêche: l'enquête qui épingle la pêche industrielle. France 3-Régions, 25 October. See also, Neslen, A. 2023. 'Phenomenal Loophole' in Quotas Could Lead to Massive Overfishing. The Guardian, 16 March.

[11] On this point, it seems difficult to obtain precise figures. Some associations mention the figure of 27 million tons each year worldwide, which in fact corresponds to an extrapolation proposed by the FAO in the early 2000s. It has since been discussed and sometimes refined to 7 million tons. See, for example, FAO. 2008. Discards from world marine fisheries. An update, XX.

[12] Roberson, L.A., Watson, R., and C.J. Klein. 2020. Over 90 Endangered Fish and Invertebrates are Caught in Industrial Fisheries. *Nature Communications* 11(4764): 1–8.

[13] Vaz, S., and P. Laffargue. 2021. Pêche: ce que la science nous dit de l'impact du chalutage sur les fonds marins. The Conversation, 15 December. Their summary links to various sources, including a report by the International Council for the Exploration of the Sea (ICES). 2021. EU Request on How Management Scenarios to Reduce Mobile Bottom Fishing Disturbance on Seafloor Habitats Affect Fisheries Landing and Value. Available at: https://ices-library.figshare.com/articles/report/EU_request_on_how_management_scenarios_to_reduce_mobile_bottom_fishing_disturbance_on_seafloor_habitats_affect_fisheries_landing_and_value/18638630; see also Jac, C., Desroy, N., Certain, G., Foveau, A.,

Lastly, a fourth direct environmental impact of industrial fishing which is less well known and less discussed, though nevertheless noteworthy, is the significant suffering such large-scale fishing methods inflict on fish and other marine animals. While this effect may seem far removed from the concerns of workers in the fishing industry (who are already subject to numerous constraints), industrial fishing methods typically show little regard for animal welfare. Industrial fishing exacerbates suffering by the sheer volume of animals involved, not only the suffering of fish (crushed and asphyxiated in nets) but also of other non-targeted animals, which are sometimes thrown back into the sea in agony, or wounded deliberately, as is the case with dolphins and sharks whose fins are sliced off by fishers when ensnared in the nets, a highly cruel measure taken to save time and avoid damaging the catching gear.[14]

Along with these considerable direct impacts caused by industrial fishing, such large-scale fishing operations have many indirect impacts on marine ecosystems, contributing to its long-term degradation. There are at least two notable types of impact to mention here. On the one hand, overfishing exacerbates the decline in marine biodiversity by causing a disruption of food chains, which, coupled with the destruction of certain habitats, further hampers the natural renewal of stocks.[15] As a result, seabird populations have declined significantly (by 70% since 1950), due to their inability to feed in sufficient quantities.[16] Additionally, the disruption of food chains, warming waters, and acidification caused by industrial fishing have all been shown to modify ocean chemistry, exacerbating the

Labrune, C., and S. Vaz. 2020. Detecting Adverse Effect on Seabed Integrity. Part 2: How Much of Seabed Habitats Are Left in Good Environmental Status by Fisheries? *Ecological Indicators* 117 (106617): 13. See also the figures and studies mentioned on the European Commission's Marine and Coastal Environment page. Available at: https://environment.ec.europa.eu/topics/marine-environment_en. Accessed 20 August 2023.

[14] See Di Concetto, A. and P. Koczorowski. 2023. The Treatment of Wild-Caught Fish Under EU Law. *European Institute for Animal Law and Policy*, Research Note 6: 28. While the scope of this chapter does not cover animal welfare issues, fishing practices and the mutilation of marine mammals, reptiles, and birds in bycatch also raise important animal welfare issues. Yet, aquatic animal welfare rules are virtually non-existent in EU law.

[15] In that sense, see the United Nations' comments in the context of the One Ocean Summit in Brest in February 2022. Source: United Nations. 2022. Oceans in danger: the main threats. https://unric.org/fr/oceans-en-danger-les-principales-menaces/. Accessed 20 August 2023.

[16] See Danger pour les oiseaux: la pêche industrielle menace de nombreuses espèces d'oiseaux marins. 2018. France Inter, 11 December.

decline in phytoplankton. The progressive decline of phytoplankton is disrupting the marine and climatic ecosystem.[17] As well as being an important source of life for marine creatures, phytoplankton are important CO_2 captors and massive oxygen producers on a global scale.

Given the massive environmental impacts of industrial fishing, the EU somewhat regulates industrial fishing practices. Under Article 3(1)(d) of the Treaty on the Functioning of the European Union (TFEU), the EU has exclusive competence for "the conservation of the biological resources of the sea within the framework of the common fisheries policy." Articles 38 and 43 TFEU define the Common Fisheries Policy (CFP) and determine the relevant procedures. Created in 1983, the CFP's initial objectives were the economic viability of the Union's fishing fleets and the preservation of fish stocks and the marine environment. However, the various reforms of the CFP, notably in 2002 and 2013, have shown that the EU's policy has failed to meet its sustainability objectives. The CFP's new regulatory framework, set out in Regulation 1380/2013, has therefore aimed to strike a better balance between the economic and social benefits of fishing and the challenges associated with ensuring its long-term sustainability.[18]

The new CFP thus relies on various tools and concepts, such as multi-annual management plans based on a so-called ecosystem approach, improved data collection, a ban on discarding unwanted catches, regular reviews of "technical measures" (e.g., the size of targeted fish, fishing gear characteristics),[19] and limits on the "maximum sustainable yield" (MSY). Achieving and adjusting the MSY—the highest possible theoretical yield in consideration of stock preservation—is the priority of the new CFP. MSY is established on the basis of scientific advice from bodies such as the International Council for the Exploration of the Sea (ICES) and the

[17] Vaz, S., and P. Laffargue. 2021. Pêche: ce que la science nous dit de l'impact du chalutage sur les fonds marins. The Conversation, 15 December.

[18] Regulation (EU) 1380/2013 of the European Parliament and of the Council of 11 December 2013 on the Common Fisheries Policy, amending Council Regulations (EC) 1954/2003 and Council Regulations (EC) 1224/2009 and repealing Council Regulations (EC) 2371/2002 and Council Regulations (EC) 639/2004 and Council Decision 2004/585/EC. A similar dynamic can be observed within the Common Agricultural Policy (CAP).

[19] On this point, see Regulation (EU) 2019/1241 of the European Parliament and of the Council of 20 June 2019 on the conservation of fisheries resources and the protection of marine ecosystems through technical measures.

Scientific, Technical and Economic Committee for Fisheries, whose members are appointed by the European Commission's Directorate-General for Maritime Affairs. This indicator then serves as the basis for the Commission's determination of fishing possibilities, i.e., "total allowable catches", then the famous "quotas" distributed among EU member states, or "fishing effort" in the Mediterranean and Black Seas (which calculate "fishing effort" through the number of days authorized at sea, as opposed to the quantity of catches).

The existing regulatory framework, however, is not sufficient to limit the harmful effects of industrial fishing on marine ecosystems. As was also highlighted in the Common Agricultural Policy,[20] there is a structural imbalance between the CFP's economic objectives and its environmental safeguards, which could jeopardize the policy's economic and environmental viability in the long term. This chapter thus provides a critical analysis of the applicable EU fishing rules, in light of existing practices and scientific analyses. The scope of this chapter will not, however, be limited solely to fish stocks targeted by fisheries and subject to quantitative and/or technical regulations. This is because the entire marine ecosystem, including all types of animals and their habitats, suffers the consequences of industrial fishing practices. The multiple, interlocking consequences of industrial fishing (stock management, habitats, unwanted catches, etc.) are difficult to separate. Indeed, all the causes and consequences mentioned are to some extent intertwined, which logically leads us to favor a global, holistic approach to regulating the interactions between industrial fishing and the marine environment.

With this holistic approach, the two main issues to remedy would be the lack of coherence in the CFP and the lack of effectiveness of the legislation in force. The CFP pursues primarily economic objectives, and although it contains elements of environmental regulation, the harmful effects of industrial fishing on marine animals and their ecosystems are still considered purely collateral by the EU Legislature. While the European Commission is at least attempting to limit the extent of these effects, it is doing so in a complex—political, economic, and social—context. While the Commission shares the findings of various scientific bodies and authorities, and is explicitly concerned about them, the Commission has struggled to involve Member States in a reform process that would achieve a

[20] See Chap. 3, Nicolas de Sadeleer and Alice Di Concetto, "The Common Agricultural Policy and Animals: Friends or Foes?"

true MSY and restore the good ecological status of the marine environment in a timely way. Such regulatory dynamics take place in the following two areas: the protection of marine habitats and ecosystems (discussed below in Sect. 2) and the reduction of incidental catches of protected marine species (discussed below in Sect. 3).

2 The Stunted Progress of Marine Habitat and Ecosystem Conservation Rules

When it comes to protecting and even restoring marine habitats and ecosystems, the EU Legislature has fallen short of integrating the CFP into a legislative environment composed of a diversity of legal instruments (discussed below in Sect. 2.1). As a result, rules regulating fishing practices remain far removed from the CFP's environmental objectives (discussed below in Sect. 2.2). At the same time, the EU Legislature's reform plans include goals with ambition levels that fall short of the expectations set by official scientific conclusions (discussed below in Sect. 2.3).

2.1 A Piecemeal Regulatory Framework

Under Article 11 TFEU, the EU is required to integrate environmental protection and sustainable development requirements into the "definition and implementation" of its policies. On this basis, and in response to the failure to ensure the long-term sustainability of European fisheries, Regulation 1380/2013 on the Common Fisheries Policy (CFP Regulation)[21] lays down ambitious environmental guiding principles and objectives. There are numerous references to these environmental principles and objectives in the recitals of the CFP Regulation. For example, in Recital 10, the EU Legislature requires that the precautionary approach should govern the sustainable exploitation of the sea's biological resources.[22] In Recital 11, the EU Legislature mandates that the CFP must "contribute to the protection of the marine environment, to the sustainable management of all commercially exploited species, and in particular to the achievement of good environmental status by 2020 as set out in Article

[21] Regulation 1380/2013 of 28 December 2013 on the Common Fisheries Policy, 2013 O.J. (L 354) 22–61.

[22] Ibid., p. 23.

1(1) of Directive 2008/56/EC."[23] The thirteenth recital further refers to the need to implement "an ecosystem-based approach to fisheries management" and to limit "environmental impacts of fishing activities," as well as to avoid and reduce "unwanted catches."[24]

Article 2 of the CFP Regulation also clearly presents the CFP's objectives. In particular, Article 2(3) states that the CFP "implement[s] the ecosystem-based approach to fisheries management so as to ensure that negative impacts of fishing activities on the marine ecosystem are minimized, and shall endeavor to ensure that aquaculture and fisheries activities avoid the degradation of the marine environment."[25]

Article 4(1)(9) of the CFP Regulation further provides a clear definition of "ecosystem-based approach to fisheries management":

> an integrated approach to managing fisheries within ecologically meaningful boundaries which seeks to manage the use of natural resources, taking account of fishing and other human activities, while preserving both the biological wealth and the biological processes necessary to safeguard the composition, structure and functioning of the habitats of the ecosystem affected, by taking into account the knowledge and uncertainties regarding biotic, abiotic and human components of ecosystems…

It is thus fair to consider that the EU Legislature had the genuine intent to make the protection of the marine environment a central pillar of the CFP. This intent is consistent with an ecosystem approach, which must permeate the various aspects of fisheries policy decided at EU level and implemented at Member State level in order for the CFP to protect marine ecosystems and ensure the sustainability of fishing activities.

Besides the CFP Regulation, other pieces of legislation relevant to fishing activities include Regulation 2016/2336 Establishing Specific Conditions for Fishing for Deep-Sea Stocks in the North-East Atlantic.[26] In its recitals, Regulation 2016/2336 refers to the environmental objectives of the CFP as well as the United Nations' call to improve the

[23] Ibid.

[24] Ibid.

[25] Ibid., p. 29.

[26] Regulation 2016/2336 of the European Parliament and of the Council of 14 December 2016 Establishing Specific Conditions for Fishing for Deep-sea Stocks in the North-East Atlantic and Provisions for Fishing in International Waters of the North-East Atlantic, 2016 O.J. (L 354) 1–19.

protection of vulnerable deep-water marine ecosystems.[27] Regulation 2016/2336 further introduces a general ban for any vessel operating in European waters, and for European vessels in international waters, on the use of bottom trawling at a depth greater than 800 meters.[28]

The Regulation also allows the European Commission to adopt and update a list of areas between 400 and 800 meters depth, to which the same protection regime must apply,[29] which the European Commission did in 2022, by way of Implementing Regulation 2022/1614.[30] This implementing regulation thus designated 87 areas hosting or likely to host vulnerable marine ecosystems, covering some 16,400 km^2, where deep-sea fishing is prohibited below 400 meters.[31] However, the scope of this regulation only covers the North-East Atlantic. Furthermore, the 87 areas designated by the European Commission for 2022, achieved through difficult political negotiations, only represent 1.16% of European waters in the region mentioned,[32] with deep-sea trawling only accounting for 1% of landings in this zone.[33] The effects of this implementing regulation toward ending deep-sea fishing are, therefore, limited.

Although they fall strictly within the scope of the CFP, these legislative acts—Regulation 2016/2336 and Implementing Regulation 2022/1614—are rooted in a legislative environment that largely encourages the protection of biodiversity and ecosystems in the EU. Outside the remit of the CFP, several other legislative acts are relevant to the

[27] Ibid., Recitals 1 and 2, p. 1.

[28] Ibid., art 8(4), p. 7.

[29] Ibid., art 9, p. 8.

[30] Commission Implementing Regulation 2022/1614 of 15 September 2022 Determining the Existing Deep-sea Fishing Areas and Establishing a List of Areas where Vulnerable Marine Ecosystems are Known to Occur or are Likely to Occur, 2022 O.J. (L 242) 1–141.

[31] Ibid., Annex I, p. 4.

[32] See, for instance, a press release from Representation in France. 2022. End of Bottom Fishing in Protected Deep-Sea Ecosystems in EU Waters. European Commission, 15 September. Available at: https://france.representation.ec.europa.eu/informations/fin-de-la-peche-de-fond-dans-les-ecosystemes-en-eau-profonde-proteges-dans-les-eaux-de-lue-2022-09-15_fr.

[33] As noted by the European Commission during the difficult debates preceding the adoption of the 2016 regulation banning trawling beyond a depth of 800 meters in the Northeast Atlantic. For more on this, see Helaoua, Y. 2013. Bruxelles renonce à interdire la pêche en eaux profonde. France 24, 10 December.

protection of marine biodiversity, starting with the Birds[34] and Habitats[35] Directives, which the EU enacted in 1979 and 1992 respectively. Both directives aim to protect endangered species and habitats within the EU, and taken together, create a network of protected areas known as "Natura 2000" sites. These two Directives cover various species and habitats considered to require particular conservation measures. Specifically, these species and habitats benefit from "special protection areas" (Birds Directive)[36] or "special areas of conservation" (Habitats Directive),[37] which form Natura 2000 sites designated and managed by the Member States. There are over 27,000 such sites in the EU, including some 3000 at sea (although Natura 2000 conservation rules do not apply to the French overseas territories). These marine Natura 2000 sites, known as "marine protected areas", cover around 450,000 km^2 of Europe's maritime zones—out of just over 5 million km^2 of exclusive economic zone under the jurisdiction of EU Member States, not including overseas territories.[38]

A second example of an EU environmental law that is relevant to fisheries is the 2008 Marine Strategy Framework Directive,[39] which established, at least in principle, common objectives for all Member States in terms of protecting the marine environment from damage caused by human activities. In particular, the Directive requires Member States to draw up strategies aimed at restoring a "good environmental status" of the marine environment by 2020, while ensuring the sustainability of the economic activities taking place there, all in consideration of precise criteria regularly updated by the Commission.[40] However, in a 2020 report, the Commission

[34] Directive 2009/147/EC of the European Parliament and of the Council of 30 November 2009 on the Conservation of Wild Birds, 2010 O.J. (L 20) 7–25.

[35] Directive 92/43/EEC of 21 May 1992 on the Conservation of Natural Habitats and of Wild Fauna and Flora, 1992 O.J. (L 206) 7–50.

[36] Directive 2009/147/EC of the European Parliament and of the Council of 30 November 2009 on the Conservation of Wild Birds, art 4(4), 2010 O.J. (L 20) 20.

[37] Directive 92/43/EEC of 21 May 1992 on the Conservation of Natural Habitats and of Wild Fauna and Flora, art 3, 1992 O.J. (L 206) 10.

[38] See the Natura 2000 Coverage in Europe's Seas page on the European Environment Agency's website at https://www.eea.europa.eu/themes/biodiversity/natura-2000/natura-2000-coverage-in-european-seas-4. Accessed 22 August 2023.

[39] Directive 2008/56/EC of the European Parliament and of the Council of 17 June 2008 Establishing a Framework for Community Action in the Field of Marine Environmental Policy, 2008 O.J. (L 164) 19–40.

[40] On this, see Commission Decision 2017/848 of 17 May 2017 Laying Down Criteria and Methodological Standards on Good Environmental Status of Marine Waters and

explicitly recognized that the Marine Strategy Framework Directive had fallen short of meeting its objectives for a number of reasons, including the lack of uniform data for the whole territory, inconsistencies in the use of recommended indicators, and persistent or even aggravated threats to marine ecosystems and fish stocks, notably due to overfishing or pollution.[41] The Marine Strategy Framework Directive was due to be reviewed before the end of 2023, but the European Commission has not yet reported on the evaluation process.

Lastly, the European Commission's Biodiversity Strategy adopted in May 2020 is also central to the protection of marine ecosystems.[42] Presented as a pillar of the European Green Deal by the European Commission, the Biodiversity Strategy presents a series of policy and regulatory actions, including the expansion of the number of protected areas, with the aim of covering at least 30% of the EU's land and marine area by 2030, as well as new restoration measures for degraded ecosystems. However, the negotiations leading to the adoption of the Nature Restoration Law have resulted in a scaled-down target of restoring 20% of land and marine areas by 2030, with the goal of addressing all ecosystems in need of restoration by 2050.[43] The Nature Restoration Law came into force on 19 August 2024,[44] and with this, Member States are required to submit their National Restoration Plans to the Commission within two years. These plans are to present the different measures Member States will implement to meet restoration goals.

There are, therefore, numerous rules, general principles, and strategies designed to protect and even restore certain marine habitats and species in EU law. However, their objectives remain far from being achieved.

Specifications and Standardised Methods for Monitoring and Assessment, 2017 O.J. (L 125) 43–74.

[41] European Commission. 2020. Report from the Commission to the European Parliament and the Council on the Implementation of the Marine Strategy Framework Directive (Directive 2008/56/EC). COM/2020/259, § 4.1–4.2.7.

[42] European Commission. 2020. Communication from the Commission to the European Parliament, the Council, the European Economic and Social Committee and the Committee of the Regions: "EU Biodiversity Strategy 2030. Bringing nature back into our lives." COM/2020/380 final.

[43] Regulation 2024/1991 of the European Parliament and of the Council of 24 June 2024 on Nature Restoration and Amending Regulation 2022/869, art 1(2), 2024 O.J. (L 2024/1991) 1–93.

[44] Ibid., art 27(1), p. 41.

2.2 *A Glaring Gap Between Stated Objectives and Actual Practice*

The ineffectiveness of conservation rules in EU law has been a major source of social and political tension. The fishing industry is already under pressure due to various factors, including changes in quotas, competition between fishing practices, the consequences of the UK's withdrawal from the EU, fleet reductions, working conditions, and energy prices. In addition to a difficult economic situation, the degradation of fish stocks and ecosystems further threatens the viability of this industry, as highlighted by numerous institutions. In two reports dating from 2020, the European Court of Auditors and the European Commission made a similar assessment of the state of fish stocks and the environmental status of ecosystems, concluding that the actions undertaken by the Union and its Member States have been insufficient to achieve adequate restoration or conservation of the marine environment.

On the one hand, such measures have fallen short of reducing overfishing, which continues unabated. Despite a slight decrease in the Atlantic, 40% of species subject to quotas there remain overfished, and that figure skyrockets to 90% in the Mediterranean.[45] In most cases, the measures implemented under the Common Agricultural Policy (CAP) do not enable MSYs to be achieved. More generally, "…the EU has not succeeded in halting the loss of marine biodiversity by 2020."[46]

On the other hand, the marine protected area mechanism, which has been widely used and promoted by the Marine Strategy Framework Directive since its adoption in 2008, offers little to no protection against the negative consequences of fisheries, particularly when trawling is involved. Additionally, the network of marine protected areas lacks clarity. There are frequent overlaps between Natura 2000 sites and areas

[45] European Court of Auditors. 2020. Marine environment: EU protection is wide but not deep. Special Report 26/2020, p. 38–39. See also, European Commission. 2020. Report from the Commission to the European Parliament and the Council on the Implementation of the Marine Strategy Framework Directive (Directive 2008/56/EC). COM/2020/259, § 4.2.2, which nevertheless highlights the progress made since only 15% of stocks were exploited within safe biological limits in 2003.

[46] European Court of Auditors. 2020. Marine environment: EU protection is wide but not deep. Special Report 26/2020, p. 37, which is based on the European Environment Agency. 2020. Marine Messages II: Navigating the Course Towards Clean, Healthy and Productive Seas Through Implementation of an Ecosystem-Based Approach. EEA Report No. 17/2019, p. 25.

benefiting from nationally defined protection; however, when an area is classified, the levels of protection are highly disparate and vary according to the species, habitats, or activities targeted. Overall, fishing is rarely prohibited or even regulated in these protected areas.

France is quite representative of these numerous shortcomings. A French Court of Auditors study of 21 Natura 2000 marine sites showed that only 14% had fishing bans in place, while 43% regulated fisheries via authorizations or technical measures, and 43% did not impose any specific regulations.[47] Similarly at the EU level, the European Environment Agency (EEA) and the European Commission have pointed out that only 1% of marine protected areas were fully protected.[48] Scientific studies have also highlighted the fact that, in many protected areas, trawling is even utilized at much higher levels than in unprotected areas.[49] Many sites actually cover traditional fishing zones close to coastlines and ports, thus resulting in inconsistencies between the scope and objectives of the applicable marine conservation rules. Lack of enforcement is also an issue, to the extent that competent authorities have used anti-spawning devices (a kind of concrete block with metal hooks to block or damage nets) in zones where fishing is prohibited.[50]

The primary causes of this alarming state of marine habitat and biodiversity degradation are overfishing and the absence of a ban on fishing, or at least on the use of towed gear, in protected areas. With these main reasons for marine habitat and biodiversity degradation, there exist other considerable flaws inherent in the existing legislative framework.

Firstly, fisheries control measures within the EU are ineffective. Member States are responsible for monitoring fishing activities in their territory, and it is the European Commission's responsibility to verify that Member States are in compliance with such monitoring.[51] However, this control

[47] Ibid., p. 27.

[48] European Environment Agency. 2020. Marine Messages II, Navigating the Course Towards Clean, Healthy and Productive Seas Through Implementation of an Ecosystem-Based Approach. EEA Report No. 17/2019, p. 31, box 3.2.

[49] Dureuil, M., Boerder, K., Burnett, K.A., Froese, R., and B. Worm. 2018. Elevated Trawling Inside Protected Areas Undermines Conservation Outcomes in a Global Fishing Hot Spot. *Science* 362(6421):1403–1407.

[50] European Court of Auditors. 2020. Marine environment: EU protection is wide but not deep. Special Report 26/2020, p. 29, Box 6, in the Cinque Terre protected area in Italy.

[51] The legal framework essentially depends on the following three regulations: Council Regulation 1005/2008 of 29 September 2008 Establishing a Community System to Prevent, Deter and Eliminate Illegal, Unreported and Unregulated Fishing; Council

mechanism has regularly been criticized.[52] A 2022 report by the European Court of Auditors provided more insights on the shortcomings of fisheries control measures,[53] pointing to the recurrence in the underreporting of catches, coupled with a significant disparity in control measures between Member States (four Member States are responsible for more than 75% of controls).[54] The European Court of Auditors also highlighted that a quarter of the infringements detected involved fishing activities in prohibited zones or continuing after quotas had been exhausted.[55] NGOs also reported regular infringements concerning trawling activities in prohibited zones.[56]

Secondly, the lack of proper articulation between the competences of the Member States and those of the EU prevents the proper implementation of EU conservation rules. Under Articles 11, 19, and 20 of the CFP Regulation (or Article 15 of the Marine Strategy Framework Directive), a Member State may adopt stricter measures to protect the marine ecosystem within its territorial sea (i.e., up to 12 nautical miles) and even beyond. However, these stricter national measures can only apply to vessels belonging to national fleets. This situation would therefore cause significant distortions in competition between national fishing operators, who must comply with stricter national legislation, and non-national fishing operators, who must only comply with minimum EU rules. A rule of this kind could even go as far as to allow non-national operators to fish in zones that would be forbidden to national operators. While Member States may submit a request to the European Commission that stricter national rules be extended to all operators fishing in national zones, the process for such a

Regulation 1224/2009 of 20 November 2009 Establishing a Community Control System for Ensuring Compliance with the Rules of the Common Fisheries Policy; Regulation 2017/2403 of the European Parliament and of the Council of 12 December 2017 on the Sustainable Management of External Fishing Fleets.

[52] For instance, see Le Figaro.fr with AFP. 2017. UE: le contrôle des pêches est insuffisant. Le Figaro, May 30; JÉGOU, B. 2021. L'Europe demande à la France des contrôles de pêche plus stricts. Le Marin-Ouest France, 4 October; Andrés, P. 2023. La révision des règles de contrôle de la pêche dans l'impasse. Euractiv, 13 March.

[53] European Court of Auditors. 2022. EU action to combat illegal fishing—Control systems in place but weakened by uneven checks and sanctions by Member States. Special Report 20/2022, p. 60.

[54] Ibid., p. 34–38.

[55] Ibid., p. 37.

[56] See all the data compiled via interactive mapping on the Med Sea Alliance campaign website. Available at: https://www.medseaalliance.org/

request is cumbersome and dissuasive. In practice, Member States do not typically adopt stricter conservation rules, with the notable exception of Portugal, which in 2014 chose to ban its national fishing fleet from trawling in a large area in and around its exclusive economic zone and continental shelf. Portugal's request to extend this measure to all fishing boats has been so slow that as of this writing the ban still only applies to Portuguese fishing operators.[57]

Thirdly, the Annexes to the Birds and Habitats Directives are outdated and do not sufficiently account for the current state of marine ecosystems, nor do they account for the endangered species listed by the International Union for Conservation of Nature (IUCN), the leading scientific organization in this field. Based on the work of the IUCN and the European Environmental Agency (EEA), the European Court of Auditors has recommended that the European Commission update the lists of areas and habitats protected by European legislation.[58] However, this recommendation does not appear to be on the Commission's agenda at present.

Lastly, existing financial support mechanisms such as the European Fund for Maritime Affairs, Fisheries and Aquaculture (FEAMP) appear out of step with current environmental needs. Over the 2014–2020 period, the FEAMP had a budget of €6 billion, of which only 6% was used to promote the development of conservation measures that could be linked to the Marine Strategy or the Birds and Habitats Directives.[59]

2.3 Limited Short-Term Prospects

The year 2022 revealed major divisions within EU institutions and agencies, on the topic of bottom trawling. In the spring of that year, the European Parliament rejected a proposal to ban trawling in protected marine areas.[60] On the occasion of the 10th anniversary of the CFP Regulation in 2023, the Commission felt the pressure from public opinion and environmental protection groups to better regulate bottom trawling, despite a political balance of power quite favorable to the fishing industry in the European Parliament and the Council. In February 2023, the

[57] European Court of Auditors. 2020. Marine environment: EU protection is wide but not deep. Special Report 26/2020, pp. 31–34.

[58] Ibid., p. 34–35.

[59] Ibid., p. 48–49.

[60] Valo, M. 2022. European Parliament Wrangles Over Fishing in Protected Areas. *Le Monde*, 5 May.

Commission thus proposed a package of measures designed to improve "the sustainability and resilience of the EU fisheries and aquaculture sector."[61] The plan was comprised of four parts, each the subject of a separate review and proposal document: a communication on energy transition in the fisheries and aquaculture sector; an action plan for the protection and restoration of marine ecosystems in support of sustainable and resilient fisheries; a communication on the Common Fisheries Policy today and tomorrow; and a report on the common organization of the markets in fishery and aquaculture products.

In these documents, the European Commission acknowledged the alarming findings of scientific authorities and explicitly questioned the long-term viability of the fisheries sector due to the various sources of pressure exerted by human activities on the marine environment and biodiversity.[62] In particular, the Commission stressed that bottom trawling was one of the most damaging activities for the seabed, and that there were still "disproportionate quantities of unwanted catches," including in protected areas.[63] In general terms, the Commission states that its Action Plan to protect and restore marine ecosystems aims to remedy the shortcomings identified by the Court of Auditors in its 2020 report, and to achieve a more coherent articulation between the EU's environmental policy and the EU's Common Fisheries Policy.[64]

However, the Commission's listed measures in the Action Plan remain somewhat vague. The Commission's main recommendation is to request that Member States progressively ban bottom fishing in marine protected areas by 2030 and to take measures to promote the use of fishing gear with a low impact on the marine environment.[65] This objective forms part of the more general framework of the EU's Biodiversity Strategy 2030, one of the aims of which is to increase the proportion of protected European waters to 30%. However, the Commission does not identify specific, binding technical measures that should be adopted to achieve

[61] European Commission. 2023. Fisheries, aquaculture and marine ecosystems: transition to clean energy and ecosystem protection for greater sustainability and resilience. Available at: https://ec.europa.eu/commission/presscorner/detail/fr/ip_23_828. Accessed 5 September 2023.

[62] European Commission. 2023. EU Action Plan: Protecting and Restoring Marine Ecosystems for Sustainable and Resilient Fisheries, p. 2–3.

[63] Ibid., p. 10.

[64] Ibid., p. 3.

[65] Ibid., p. 6–8 and 11.

this. Instead, the European Commission has set out a timeline for Member States to submit an initial roadmap by the spring of 2024, currently monitored by the European Commission, who will also support the development of innovative solutions via the FEAMP and LIFE funding programs.[66] In addition to the issue of marine protected areas, the Commission considers the need to identify and implement "threshold values" (by way of indicators yet to be defined) aimed at protecting the integrity of the seabed.[67] The Nature Restoration Law refers to these threshold values, although their specifications are yet to be determined.

Although the Action Plan illustrates the Commission's intention to introduce a more environmentally friendly approach to the CFP, the European Commission has not committed to adopting any binding measures that would result in the revision of existing regulations (not for the CFP Regulation, or for Regulation 2019/1241 on the Conservation of Fisheries Resources and Protection of Marine Ecosystems Through Technical Measures, which had already been amended in 2019). The Action Plan also appears relatively unambitious in light of the scientific findings that had prompted its adoption. The European Commission only considers a prohibition on trawling in protected sites and in a way that must not undermine the competitiveness of the fishing industry. Yet, studies have shown that banning trawling in protected areas simply shifts the fishing effort to other, unprotected areas, which are then subject to increased ecological degradation.

Ultimately, the prospects for changes to the regulatory environment of the CFP, and the protection of marine habitats and ecosystems, remain limited. The current state of proposals, the policy orientations envisioned by the European Commission, and discussions between stakeholders mainly revealed the tension surrounding plans to revise the CFP. Whether the targets set for 2030 are as unrealistic as those initially set for 2020 under the Marine Strategy is therefore still unclear. However, beyond the state of the marine ecosystem, the shortcomings in the EU's fisheries policy orientations threaten the very viability of the fisheries sector, with significant political risks for the European Commission and the Member States, should these institutions fail to remedy the situation.

[66] Ibid., p. 13.
[67] Ibid., p. 11.

3 Limiting the Accidental Capture of Protected Marine Species: A Legislative and Technical Gridlock?

Accidental catches of protected marine species are significant enough to endanger the renewal of certain species. Yet, the ways in which the EU Legislature addresses this issue reveals a structural lack of policy coherence, and a lack of political will as with the issues mentioned in the previous section. The fisheries sector (at least its most industrialized segments) finds itself at a technical roadblock, insofar as the most efficient fishing methods increase the risk of bycatch. Under these conditions, the EU Legislature tolerates such catches as long as operators can show they are unavoidable and unintentional (discussed below in Sect. 3.1). Despite numerous warnings from various scientific, political, and community institutions, the EU Legislature is considering the adoption of mitigation measures, and in some cases has implemented such measures. However, for the time being, these measures remain largely insufficient to offset the harmful consequences of accidental catches (discussed below in Sect. 3.2).

3.1 An EU Legal Regime Tolerant of Unwanted Catches

The EU legislation on fisheries includes several provisions that aim to limit, and sometimes reduce, "unwanted catches." This term is not explicitly defined in the CFP Regulation 1380/2013, or in Regulation 2019/1241 on the Conservation of Fisheries Resources and the Protection of Marine Ecosystems Through Technical Measures (Regulation on Technical Measures). However, it is possible to presume from these texts that, in practice, the expression "unwanted catches" refers to two types of catches regulated under these two statutes: "bycatches" and "incidental catches."

Bycatches are unintentional catches of animals belonging to species that are subject to quotas, but whose catch is not allowed due to other factors. For example, a catch from an authorized species is considered bycatch when the size of the animal is too small, so as to limit the capture of juveniles. Bycatch is not prohibited under the CFP Regulation, but operators must avoid them and must comply with a landing obligation, in order to avoid discards at sea.[68] Operators must further register bycatches, and if necessary, count them against quotas, before they can use such catches to manufacture feed for aquaculture, for instance. This rather ambivalent solution aims to avoid any discarding at sea, which would ultimately contribute to wasting otherwise perfectly consumable fish. However, these bycatches cannot be marketed directly to consumers, so as to limit their profitability and thereby disincentivize bycatch.

"Incidental catches," on the other hand, refers to the capture of protected marine animals. The intentional capture of these animals is thus prohibited. Therefore, operators must not land them if the animal is still alive or in a condition to survive. Current regulations are designed to limit such accidental catches as much as possible, although the scope of these regulations is rather limited.

While the 2013 CFP Regulation sets out general objectives for reducing all types of unwanted catches, Regulation 2019/1241 on the Conservation of Fisheries Resources and the Protection of Marine Ecosystems Through Technical Measures specifically deals with the incidental capture of protected animals.[69] Article 3(2)(b) thus provides that technical measures must contribute to "[ensuring] that incidental catches of sensitive marine species, including those listed under Directives 92/43/EEC and 2009/147/EC, that are a result of fishing, are minimized and where possible eliminated so that they do not represent a threat to the conservation status of these species."[70] Similarly, Article 4(1)(b) states that technical measures must ensure that "incidental catches of marine mammals, marine reptiles, seabirds and other non-commercially exploited species do not exceed levels provided for in Union legislation and international agreements that are binding on the Union."[71]

The Regulation on Technical Measures thus ensures coherence between the CFP and environmental legislation, specifically the Habitats and Birds Directives. Section 3 of the Regulation on Technical Measures specifically covers the protection of sensitive species and habitats in a way that is consistent with the Habitats and Birds Directive. Article 10 of the Regulation

[68] Regulation 1380/2013 of 28 December 2013 on the Common Fisheries Policy, art 15, 2013 O.J. (L 354) 33–37.

[69] Regulation 2019/1241 of the European Parliament and of the Council of 20 June 2019 on the Conservation of Fisheries Resources and the Protection of Marine Ecosystems Through Technical Measures, 2019 O.J. (L 198) 105–201.

[70] Ibid., art 3(2)(b), p. 112.

[71] Ibid., art 4(1)(b).

on Technical Measures thus prohibits "the catching, retention on board, transshipment or landing of fish or shellfish species referred to in Annex IV to Directive 92/43/EEC,"[72] in addition to the species listed in Annex I(2) to the Regulation on Technical Measures.[73] Article 11 also contains a prohibition on "the catching, retention on board, transshipment or landing of marine mammals or marine reptiles referred to in Annexes II and IV to Directive 92/43/EEC and of species of seabirds covered by Directive 2009/147/EC."[74] This provision is consistent with Article 12 of the Habitats Directive, which prohibits "all forms of deliberate capture or killing of" listed in Annex IV(a), which include, for instance, all species of cetaceans.[75] Article 11 of Regulation on the Technical Measures is also consistent with Article 5 of the Birds Directive, which prohibits the "deliberate killing or capture" of bird species that are not covered by exemptions.[76]

However, the overall regime on the prohibition of unwanted catches leaves some latitude to fisheries operators, especially regarding bycatch. For example, Recital 31 of the CFP Regulation presents unwanted catches as "unavoidable even when all the measures for their reduction are applied."[77] Generally speaking, the measures contained in the law simply aim to reduce "as far as possible unwanted catches."[78] Even the Habitats and Birds Directives only prohibit the "intentional" capture of protected animals. This prohibition is further eroded in the Regulation on Technical Measures, which provides that animals from protected species who were accidentally caught must "not be harmed and [...] promptly released."[79]

As a result, the prohibition on catching protected animal species has a particularly narrow scope in the context of commercial fishing activities. This is because the link between fishing activities regulated by the CFP and marine animal protection regimes relies on the idea that unwanted

[72] Ibid., art 10(1), p. 120.

[73] Ibid., art 10(2).

[74] Ibid., art 11(1).

[75] Directive 92/43/EEC of 21 May 1992 on the Conservation of Natural Habitats and of Wild Fauna and Flora, art 12, 1992 O.J. (L 206) 7–50.

[76] Directive 2009/147/EC of the European Parliament and of the Council of 30 November 2009 on the Conservation of Wild Birds, art 5, 2010 O.J. (L 20) 9.

[77] Regulation 1380/2013 of 28 December 2013 on the Common Fisheries Policy, Recital 31, 2013 O.J. (L 354).

[78] Ibid., art 2(5)(a), p. 29.

[79] Regulation 2019/1241 of the European Parliament and of the Council of 20 June 2019 on the Conservation of Fisheries Resources and the Protection of Marine Ecosystems Through Technical Measures, art 10(3) and 11(2), 2019 O.J. (L 198) 21.

catches are accidental. In other words, the Legislature considers unwanted catches as collateral damage inherent to the activity of fishing. Yet, it is both the increase in the fishing activity and the evolution of fishing techniques that create the conditions for large numbers of protected marine animals to be ripped from their habitats. Artisanal fishing that uses methods with comparatively lower environmental impact (such as traps or trolling lines) limits catches of this type, while the use of large trawls makes accidental catches almost inevitable and increases the likelihood of injuring or killing the animals concerned. The EU Legislature therefore improperly qualifies these catches as "accidental" since fishing practices are evolving in such a way that they are becoming the norm.

3.2 Ineffective Implementation of Preventive Measures in National Law: France as a Case Study

Member States are ultimately responsible for ensuring the proper implementation of EU law, including conservation measures in the specific context of commercial fishing activities, to the extent that unwanted catches take place in national waters. However, the implementation measures adopted in national law often fall short of achieving the objectives set in EU law, as illustrated by the increase in the numbers of animals caught as bycatch, which have continued unabated over the past three decades.

There is scientific consensus over the fact that unwanted catches have increased over the past three decades. Although it is hard to account for the volume of animals killed in the context of unwanted catches in the world, a 2008 FAO report estimated that discards accounted for 7.3 million tons per year over the period 1992–2001.[80] Given the proliferation of industrial fishing over the past two decades, this figure has undoubtedly risen sharply. Some NGOs even estimate that unwanted catches account for 27 million tons per year, but this estimate has yet to be officially verified.[81] Towed gear is responsible for a large proportion of these catches, since at least 20 to 30% of the contents of towed gear does not correspond

[80] FAO. 2008. Fisheries Technical Paper. No. 470: Discards in the World's Marine Fisheries - An Update. Available at: https://www.fao.org/3/y5936f/y5936f00.htm

[81] See, for example, the WWF website page on Fighting bycatch at sea. Available at: https://www.wwf.fr/champs-daction/ocean/peche-aquaculture/prises-accessoires. Accessed 20 September 2023.

to the target species, as the European Commission pointed out in its 2023 report on the CFP.[82]

More specific data is available regarding the bycatch of protected animals, especially cetaceans at national level. In France, the Institut Pelagis and the *Réseau National Échouages* calculates that, between January and April of 2019, 1231 cetaceans were found stranded on the French Atlantic coast, 85% of whom "showed internal and external traces of death in fishing gear."[83] The stranding of cetaceans as bycatch is relatively constant in the Bay of Biscay. For example, 1067 cetacean strandings were recorded on the Atlantic coast in winter 2020 and 1143 in winter 2021.[84] Because only a small number of these mammals are stranded on beaches, researchers estimate that around 9000 cetaceans are caught as bycatch annually in the Bay of Biscay alone.[85] Moreover, incidental catches have increased sharply from 1990 to 2020: around 89,000 dolphins (by far the cetaceans most affected by incidental catches) were affected in the years between 1990 and 2019, with the annual average more than doubling to 7800 each year between 2016 and 2019.[86] Furthermore, based on stranding data, and data on fishing efforts and the navigation of vessels over 12 meters, a 2021 scientific study established a correlation between the increase in fishing activities, the use of towed gear, and the increase in bycatch.[87] The European Commission shared similar observations in its

[82] European Commission. 2023. The Common Fisheries Policy Today and Tomorrow: A Fisheries and Oceans Pact Towards Sustainable, Science-Based, Innovative and Inclusive Fisheries Management, p. 10.

[83] Peltier, H., Authier, M., Caurant, F., Dabin, W., Daniel, P., Dars, C., Demaret, F., Meheust, E., Van Canneyt, O., Spitz, J., and V. Ridoux. 2019. État des connaissances sur les captures accidentelles de dauphins communs dans le golfe de Gascogne - Synthèse 2019. Rapport scientifique dans le cadre de la convention avec le MTES. Observatoire PELAGIS - UMS 3462. La Rochelle Université/CNRS, p. 6.

[84] Dars, C., Méheust, E., Genu, M., Méndez-Fernandez, P., Peltier, H., Wund, S., Caurant, F., Dabin, W., Demaret, F., Spitz, J., and O. Van Canneyt. 2021. Les échouages de mammifères marins sur le littoral français en 2021. Rapport scientifique de l'Observatoire Pelagis. La Rochelle University and CNRS, p. 12.

[85] ICES. 2023. EU Additional Request on Mitigation Measures to Reduce Bycatches of Common Dolphin (Delphinus Delphis) in the Bay of Biscay and Iberian Coast. ICES Special Request Advice. sr.2023.01, p. 11.

[86] Dars, C., Méheust, E., Genu, M., Méndez-Fernandez, P., Peltier, H., Wund, S., Caurant, F., Dabin, W., Demaret, F., Spitz, J., and O. Van Canneyt. 2021. Les échouages de mammifères marins sur le littoral français en 2021. Rapport scientifique de l'Observatoire Pelagis. La Rochelle University and CNRS, p. 10.

[87] Ibid., p. 12.

2023 Action Plan for the Protection and Restoration of Marine Ecosystems, which states that "for some [marine animals], the incidental capture in fishing gear is one of the main threats. These problems have contributed to the decline of populations of several species, and although most are strictly protected by EU nature legislation, some are still at risk of extinction."[88]

The European Commission has repeatedly criticized national competent authorities for failing to comply with their obligations to protect marine mammals species from bycatch. Over the past five years, European scientific authorities have recommended that several key EU Member States, including France, step up the fight against marine mammal bycatch. In 2020, a report by the International Council for the Exploration of the Sea (ICES), the advisory body to the European Commission in the design and implementation of the CFP, recommended the adoption of drastic measures to limit cetacean bycatch in the Bay of Biscay in France, including temporary fishing bans to allow cetacean populations to reproduce.[89] Against this backdrop, the European Commission launched a series of infringement proceedings in July 2020 against France, Spain, and Sweden, by way of a formal notice (the first of three steps in an infringement procedure). On this occasion, the Commission referred to "well-documented evidence of these species [dolphins and harbour porpoises] being caught in fishing gear"[90] and urged the three Member States to take action to reduce bycatch in their waters, in line with their obligations under the Habitats Directive and the CFP Regulation.

In 2022, the European Commission further sent a reasoned opinion (the second of three steps in an infringement procedure) to France and Spain, in addition to a letter of formal notice to Bulgaria. In doing so, the European Commission urged these Member States to take urgent action to comply with current legislation. Specifically in the case of France, the Commission highlighted three types of shortcomings in its reasoned opinion: the lack of measures to improve monitoring of the conservation status of several species and to control catches; the failure to fully transpose the Habitats Directive with regard to bycatch and conservation measures; and

[88] European Commission. 2023. EU Action Plan: Protecting and Restoring Marine Ecosystems for Sustainable and Resilient Fisheries, p. 4.

[89] ICES. 2020. EU Request on Emergency Measures to Prevent Bycatch of Common Dolphin (Delphinus delphis) and Baltic Proper Harbour Porpoise (Phocoena Phocoena) in the Northeast Atlantic. ICES Special Request Advice. sr.2020.04.

[90] Infringement Procedure INFR(2020)4036.

the absence of effective measures to reduce bycatch, such as the establishment of closed seasons and areas for fishing activities in the Bay of Biscay.[91]

The implementation of EU conservation rules in the context of commercial fishing in France has been particularly contentious, thereby illustrating the challenges of implementing EU law into national law. Overall, the French competent authorities have not displayed a sense of urgency in addressing the Commission's concerns over bycatch in French waters, likely on account of their reluctance to inflame relations with a fishing sector already under strain. The French administration first adopted a decree in 2011, establishing a list of marine mammals protected on national territory and the terms and conditions of their protection. However, this measure only took effect in 2019, following the entry into force of another decree requiring fisheries operators to declare the capture of marine mammals—electronically for vessels over 12 meters in length, and via paper fishing log for smaller vessels—so as to monitor the number of bycaught animals.[92]

With regard to prevention and reduction measures, an executive order from 2019 made it compulsory for fishing operators to install active acoustic repulsion devices (pingers) on pelagic trawlers over 12 meters long in the Bay of Biscay[93] from January to April each year. A 2020 executive order later extended the scope of this obligation to apply year-round, to all pelagic trawlers, of any size, as well as demersal trawlers in pairs through a 2020 executive order.[94] More recently, a 2022 decree made it compulsory for all vessels over 6 meters in length fishing in the Bay of Biscay to carry Vessel Monitoring System beacons, and included vessels over 15 meters in an on-board observation program with the aim of documenting accidental catches.[95] Another decree, also adopted in 2022, introduced a number of

[91] Ibid. A summary of the reasoned opinion is available at: https://ec.europa.eu/commission/presscorner/detail/FR/inf_22_3768. Accessed 20 September 2023.

[92] Arrêté du 6 septembre 2018 portant modification de l'arrêté du 1er juillet 2011 fixant la liste des mammifères marins protégés sur le territoire national et les modalités de leur protection.

[93] Arrêté du 26 décembre 2019 portant obligation d'équipement de dispositifs de dissuasion acoustique pour les chaluts pélagiques dans le golfe de Gascogne.

[94] Arrêté du 27 novembre 2020 portant modification de l'arrêté du 26 décembre 2019 portant obligation d'équipement de dispositifs de dissuasion acoustique pour les chaluts pélagiques dans le golfe de Gascogne.

[95] Arrêté du 27 décembre 2022 relatif à l'obligation d'emport d'équipements du système de surveillance des navires embarqués à bord de navires de pêche de moins de douze mètres sous pavillon français.

pilot programs running from 2023 to 2025, including the use of acoustic repellents by vessels using gillnets and entangling nets, and the installation of on-board cameras on approximately 100 vessels operating in the Bay of Biscay.[96] The French competent authorities presented these series of prevention and reduction measures as part of an Action Plan aimed at remedying the shortcomings identified in the European Commission's infringement procedure.[97]

However, many have expressed their frustration over this Action Plan, starting with environmental protection advocates. In a 2023 ruling, the *Conseil d'État*—the highest administrative court in France—ultimately sided with environmental advocates in a case against the French administration, by deciding that French competent authorities had failed to take effective measures to curb the accidental killing of cetaceans.[98] In its decision, the Court first noted the increasing number of studies demonstrating the alarming reduction in cetacean populations in the Bay of Biscay.[99] The Court also pointed out that measures based on the use of acoustic repellents were insufficient, due to their ineffectiveness, but also because of their counterproductive effects, such as causing small cetaceans to flee their habitat and reducing their chances of survival. As a result, the Court stated, the use of acoustic devices had to be combined with measures to reduce fishing efforts, such as fishing bans.[100]

The *Conseil d'Etat* therefore concluded that current fishing activities in the Bay of Biscay constitute a serious threat to the conservation of small cetaceans.[101] Secondly, the *Conseil d'Eta*t pointed to the lack of measures that would seek to provide precise data on the scale and causes of bycatch in the Bay of Biscay. The Conseil specifically pointed out that fishing operators declared underestimated figures as part of their reporting obligations (only 69 catches were declared during the winter of 2021), suggesting the

[96] Arrêté du 29 décembre 2022 relatif à l'amélioration de la collecte de données sur les captures accidentelles d'espèces protégées et à l'expérimentation de dispositifs techniques de réduction des captures accidentelles de dauphin commun à bord de navires de pêche sous pavillon français.

[97] Secrétariat d'État chargé de la mer. 2023. Les engagements de l'État, des pêcheurs et des scientifiques pour lutter contre les captures accidentelles de petits cétacés en Atlantique. https://mer.gouv.fr/cetaces. Accessed 20 September 2023.

[98] Conseil d'État. 2023. France Nature Environnement, Défense des milieux aquatiques, Sea Shepherd France. Nos 449788, 449849, 453700, 459153.

[99] Ibid., § 28.

[100] Ibid., § 29.

[101] Ibid., § 33.

existence of significant enforcement gaps due either to fishing operators' lack of compliance with reporting requirements, or to the ineffectiveness of observers or on-board cameras.[102] The Conseil d'Etat thus ordered the French administration to implement appropriate spatial and temporal fishing bans, in addition to adopting additional measures so as to allow the proper monitoring of catches and the evaluation of conservation measures.[103]

The ruling of the *Conseil d'Etat* constitutes a significant rebuke to the French competent authorities, whose approach had relied on adopting non-binding or experimental measures that have not been adequate to curb incidental catches. The French government thus appears to be caught in a bind between a struggling fishing industry that would consider even a partial fishing ban as a political abandonment (undoubtedly precipitating the decline or bankruptcy of certain players), and the European Commission and environmental groups, which are likely to attain their goals through litigation, as suggested by the recent ruling by the *Conseil d'Etat*. The European Commission may consider moving forward with the infringement procedure against France, as the Commission is likely to derive great political benefit from appearing as the protector of nature.

The risk that the French competent authorities might be challenged before the courts is thus relatively high, even more when considering the lack of reduction in accidental catches. In 2023, French competent authorities thus reported inconclusive figures on the effectiveness of the measures adopted as a result of the Action Plan. Over a period of four to five months in 2022–2023, 1482 cetaceans have been found stranded on the Atlantic coast alone. Over the same period, fishing vessels have reported 238 incidental catches.[104] These numbers show that the compliance rate with the reporting obligations is slightly better than in previous years. However, the reported number of incidental catches is still very low compared to the number of strandings recorded by researchers, which is the basis they use to estimate a much higher total number of catches. With regard to vessels equipped with on-board cameras, the French competent authorities also reported that seven out of the 20 participating vessels had

[102] Ibid., § 38.

[103] Ibid., § 40–41.

[104] Ministère de la Mer. 2023. Capture de cétacés: Hiver 2022–2023. Bulletin n°10—Du 16 au 30 avril, p. 4. Available at: https://mer.gouv.fr/sites/default/files/2023-05/20230509_Bulletin_n10.pdf

dropped out of the pilot program by disconnecting their on-board systems in February/March 2023.[105]

In May of 2023, France announced it would implement additional prevention measures for another year, during the winter of 2023–2024, before resorting to fishing bans in the Bay of Biscay as ordered by the *Conseil d'État* in its 2023 ruling. These additional prevention measures include a control plan to ensure vessels are equipped with pingers, as well as programs to observe fisheries, analyze risks (particularly in Natura 2000 sites), and raise awareness among fishing operators.[106] In October 2023, the French competent authorities subsequently changed their approach and issued, by way of a decree, a fishing ban in the Bay of Biscay for the months of January and February 2024, 2025, and 2026 for vessels of eight meters or more equipped with specific nets.[107] This change was likely motivated by the risk of litigation that the competent authorities faced. This decree also included a number of transitional, technical exemptions. For instance, vessels equipped with catch reduction devices were exempted from the scope of the decree, as well as vessels using certain fishing gear, such as pelagic seines. However, the *Conseil d'Etat* issued an order in December 2023, suspending the application of the various derogations.[108]

Similarly to the issues relating to the protection of marine habitats and ecosystems discussed in the first section of this chapter, the absence of tangible progress in preventing and reducing bycatch is bound to lead to more legal and political tensions. Such tensions are rooted in the conflicting economic and environmental objectives of the CFP. The accumulation of technical measures with limited scope is unlikely to lead to a resolution of this structural conflict. Until EU policymakers put adequate environmental and animal protective measures in place, the EU fisheries sector will continue to endure various crises that jeopardize the viability of its activities.

[105] Ibid., p. 6.

[106] Secrétariat d'État chargé de la mer. 2023. Les engagements de l'État, des pêcheurs et des scientifiques pour lutter contre les captures accidentelles de petits cétacés en Atlantique. Available at: https://mer.gouv.fr/cetaces. Accessed 20 September 2023.

[107] Arrêté du 24 octobre 2023 établissant des mesures spatio-temporelles visant la réduction des captures accidentelles de petits cétacés dans le golfe de Gascogne pour les années 2024, 2025 et 2026.

[108] Conseil d'Etat. 2023. France Nature Environnement, Ligue pour la protection des oiseaux, Sea Shepherd France. Nos 489,926, 489,932, 489,949.

Cruelty Under Contract: A Case Study in French Law

Ilyana Aït-Ahmed and Solène Kerisit

1 Introduction

In 2020, half of French farmers reported an annual income below €22,500, making farmers one of the most socioeconomically vulnerable groups in the country. Some 17.7% lived below the poverty line, compared to 14.4%

Acknowledgement: The title and the opening section of this chapter draw on unpublished works by Alice Di Concetto, whom the authors thank for kindly sharing her findings for this chapter. The authors are also grateful to Aude-Solveig Epstein for her significant contributions throughout the manuscript. The core of this article was drafted in 2023 and stems from a presentation given by the authors on "The Legal Framework of Farmers Disempowerment" as part of the symposium on "Animals in Economic Law in the UE".

I. Aït-Ahmed (✉)
Lawyer at the Paris Bar, Paris, France
e-mail: ilyana.aitahmed@sciencespo.fr

S. Kerisit
Québec Bar School, Montréal, QC, Canada
e-mail: solene.kerisit@sciencespo.fr

A.-S. Epstein, A. Di Concetto (eds.), *Animals in EU Economic Law*, The Palgrave Macmillan Animal Ethics Series,
https://doi.org/10.1007/978-3-032-16171-0_5

of the general population of France. Animal agriculture is particularly overrepresented in the poorest households: in 2018, one quarter of cattle, goat, and sheep farmers were below the poverty threshold.[1] French farmers also experience disproportionately poor mental health. On average, one farmer dies by suicide every day,[2] and the overall mortality rate for farmers is abnormally high compared to the rest of the population.[3] This elevated death rate is partly linked to the precariousness of farmers' livelihoods and unusually high levels of burnout, as between 2011 and 2019, 35.5% of French farmers reported being in a state of burnout.[4]

This dire situation is one of the many deleterious consequences of factory farming. The deterioration of farmers' well-being coincides with the concentration and consolidation in EU animal agriculture—trends characteristic of industrialization. In 2020, continental France accounted for nearly 390,000 agricultural holdings, representing a decline of 100,000 compared to 2010.[5] Concurrently, since the 1950s, animal production volumes have nearly quintupled,[6] while the number of animals per holding has grown sharply[7] due to the industrialization of animal farming. Even

[1] INSEE. 2021. *Le niveau de vie des ménages agricoles est plus faible dans les territoires d'élevage*. Insee Première, no. 1876. Available at: https://www.insee.fr/fr/statistiques/5434584.

[2] Mutuelle sociale agricole. 2022. Étude—Les statistiques de la MSA, La mortalité par suicide au régime agricole dans le Système national des données de santé (SNDS). Available at: https://statistiques.msa.fr/wp-content/uploads/2022/10/Etude-mortalite-par-suicide_ok.pdf.

[3] Khireddine-Medouni, I., Breuillard, É., and C. Bossard. 2016. Surveillance de la mortalité par suicide des agriculteurs exploitants. Situation 2010–2011 et évolution 2007–2011. Santé publique France. Available at: https://www.santepubliquefrance.fr/maladies-et-traumatismes/sante-mentale/suicides-et-tentatives-de-suicide/documents/rapport-synthese/surveillance-de-la-mortalite-par-suicide-des-agriculteurs-exploitants-situation-2010-2011-et-evolution-2007-2011.

[4] Torrès, O. and C. Kinowski-Moysan. 2019. Dépistage de l'épuisement et prévention du burnout des dirigeants de PME: d'une recherche académique à une valorisation sociétale. *Revue française de gestion* 284: 171–189. Available at: https://shs.cairn.info/article/RFG_284_0171?lang=fr.

[5] INSEE. 2024. Transformations de l'agriculture et des consommations alimentaires. Insee Références. Available at: https://www.insee.fr/fr/statistiques/fichier/7728903/AGRI24.pdf.

[6] Ritchie, H. and M. Roser. 2017. Meat and Seafood Production and Consumption. Our World in Data. Available at: https://ourworldindata.org/meat-and-seafood-production-consumption. Accessed 2 July 2025.

[7] Cour des comptes. 2023. Les soutiens publics aux éleveurs bovins. Available at: https://www.ccomptes.fr/fr/publications/les-soutiens-publics-aux-eleveurs-de-bovins.

though there is no explicit, standardized regulatory definition of industrial farm animal production in national or EU law,[8] it is estimated that 80% of farmed animals are raised under industrial conditions, i.e., on farms with cages and/or high stocking densities and lack of outdoor access.[9] It is further estimated that the EU is home to over 516 million chickens and pigs on approximately 22,263 industrial farms, making industrial animal farming the dominant model of food production in the EU.[10]

Industrial farm animal production is commonly characterized by methods of production that generate high production volumes on limited land, creating many externalities, including environmental harms.[11] Various criteria could serve as a basis for a definition of industrial farming, such as animal welfare standards (use of extreme confinement systems, high stocking densities, routine mutilations) or environmental standards (emission levels, size of farms).[12] Another critical criterion, however, is its economic structure, characterized by vertical integration and a heavy reliance on contract farming, especially integration contracts.[13]

Contract farming has been analyzed in US legal scholarship from the perspective of antitrust and competition law and policy,[14] but this topic

[8] Di Concetto, A. 2025. Les productions animales intensives en droit: état des lieux et perspectives de transitions. *Revue Juridique de l'Environnement.*

[9] Finger, S. 2018. Porcs, bovins, volailles, la vraie vie de la viande française dans la production animale intensive. Libération, 4 June. Available at: https://www.liberation.fr/france/2018/06/04/porcs-bovins-volailles-la-vraie-vie-de-la-viande-francaise-dans-l-elevage-intensif_1656528/; L214 Éthiques et Animaux. 2025. Key figures on animal suffering. Available at: https://www.l214.com/animaux/chiffres-cles/statistiques-pourcentage-elevage-intensif-viande-lait-oeufs/.

[10] AGtivist Agency. 2025. Megafarm Europe. Available at: https://stories.agtivistagency.com/megafarm-europe/.

[11] Charvet, J.-P. 2014. *Élevage.* Encyclopædia Universalis. Available at: http://www.universalis-edu.com/encyclopedie/elevage/; Porcher, J. 2007. Misère des productions animales. Textes et documents pour la classe. N° 939 – CNDP-CRDP, p. 18–19; European Environment Agency Glossary. Intensive Farming. Available at: https://www.eea.europa.eu/help/glossary/eea-glossary/intensive-farming.

[12] Di Concetto, A. 2025. Les productions animales intensives en droit: état des lieux et perspectives de transitions. *Revue Juridique de l'Environnement.*

[13] Ibid.

[14] For instance, see the work of Austin Frerick at Yale University and the Thurman Arnold Project on agriculture and antitrust at Yale University. Available at: https://som.yale.edu/centers/thurman-arnold-project-at-yale/agriculture-and-antitrust; Muraskin, D.S. 2025. Antitrust and Animal Agriculture. In *Elgar Concise Encyclopedia of Animal Law*, ed. Joan E. Schaffner, 55–58. Cheltenham: Edward Elgar Publishing. Available at: https://www.elgaronline.com/display/book/9781803923673/chapter15.xml.

deserves more attention from Europeans, as well as from animal law and policy scholars. Moreover, both media debates and scholarly discussions around factory farming often treat animal and farmer interests as separate and often opposing issues. As a result, the interdependence between protecting animals and farmers remains underexplored in academic literature and EU policies. Yet, farmers' freedom to choose their economic model is central to transitioning toward more humane food systems, as farmers should be able to freely engage in humane farming, if they so decide, or to abstain from animal agriculture altogether.[15]

Through the lens of animal law and policy and taking France as a case study, this chapter develops a novel account of how contract farming harms both farmers and animals, demonstrating how these harms are often mutually reinforcing.[16] It first situates contract farming within French law, highlighting the inconsistent and ultimately inadequate nature of the existing regulatory framework. It then examines a range of contractual practices that operate as vectors of exploitation for both farmers and animals, before turning to the legislative responses adopted to curb these abuses – and the reasons why they have largely failed to do so.

2 Contract Farming in French Law: An Inconsistent and Inadequate Regulatory Framework

The media often present the business model of large food retail groups, which operate on tight profit margins, as the main cause of farmers' precarity in animal agriculture.[17] According to this idea, retailers pressure farmers to lower production costs in order to offer food at competitive prices while still making a profit. While partly accurate, this explanation tends to reduce the exploitation of farmers and animals to supposedly natural market dynamics, obscuring the regulatory framework that enables such exploitation in the first place. A closer analysis of the rules governing

[15] Garcès, L. 2024. *Transfarmation: The Movement to Free Us from Factory Farming*. Boston: Beacon Press.

[16] This analysis is based on a review of integration and cooperative contracts that are covered by confidentiality agreements.

[17] Pistorius, M. 2021. 'Price war' Between French Farmers and Big Supermarkets Continues. Euractiv, 4 March. Available at: https://www.euractiv.com/section/agriculture-food/news/price-war-between-french-farmers-and-big-supermarkets-continues/.

the contractual relations between producers and processors shows that exploitative dynamics take place at a much earlier stage in the value chain, specifically in the contractual practices between farmers and integrators.

These business practices are economically structured by vertical integration and contract farming (Sect. 2.1). Legally, French law establishes a strict distinction between two types of vertical integration contracts, depending on whether contract farmers are integrated into the network of a producers' cooperative or that of a capitalist firm (referred to here as a "corporation") (Sect. 2.2).

2.1 Vertical Integration and Contract Farming

Vertical integration in animal agriculture—where a grower is tied to the production chain (typically to a processor) through contract—follows a model first developed in US poultry production in the 1960s.[18] In this contract-based model, farmers provide labor, housing, utilities, on-farm structures, and equipment, while integrators supply inputs such as young animals, feed, and medication.[19] Each contracting party retains ownership of their inputs, meaning that growers own the land, facilities, and equipment, while integrators retain ownership of farmed animals throughout the production cycle.[20]

Contract farming came to prominence in the US following the Great Depression, as a more industrialized agricultural economy emerged. Although contracts were already in use by the early 1900s, their

[18] USDA, Economic Research Service. 1996. Farmers' Use of Marketing and Production Contracts. Available at: https://www.ers.usda.gov/webdocs/publications/40764/18614_aer747a_1_.pdf?v=0.

[19] MacDonald, J., Perry, J., Ahearn, M., Banker, D., Chambers, W., Dimitri, C., Key, N., Nelson, K., and L. Southard. 2004. Contracts, Markets and Prices: Organizing the Production and Use of Agricultural Commodities. USDA, ERS, Agricultural Economic Report No. 837. Available at: https://www.ers.usda.gov/webdocs/publications/aer837/14700_aer837_1_.pdf?v=41061.

[20] Viinikainen, T. and C. Bullón Caro. 2018. Enabling Regulatory Frameworks for Contract Farming. Food and Agriculture Organization of the United Nations, p. 1–19. Available at: https://openknowledge.fao.org/server/api/core/bitstreams/39f900a0-567f-4a35-bc6b-5b3756f52b14/content; World Bank Group. 2014. An Analytical Toolkit for Support to Contract Farming. Available at: https://documents1.worldbank.org/curated/en/575871468204575206/pdf/881810REVISED00y020140WB0Internal02.pdf.

prevalence expanded significantly in the 1960s, beginning with poultry,[21] which provided the blueprint for contract farming in animal agriculture around the world, including in France. This model later extended to pork production in the late 1980s and gradually spread across virtually all sectors of animal agriculture by the early 2000s.

Contract farming has captured so much of the animal agriculture industry because it initially appeared as though it would be a boon to farmers. Marketing contracts with integrators were presented as a comfortable arrangement in which farmers would feel integrated into the larger food production process, rather than bearing alone the most risky segment of the production chain—the rearing of live animals—in complete isolation from the other operators in the chain. Farmers often chose to enter into contracts with processors because of the seeming stability such a contract would provide. With prices determined in advance, a farmer would no longer have to go through the onerous tasks of public auction and price discovery and could therefore dedicate more time to farming while enjoying a guaranteed minimum income. This advantage has been especially appealing to growers of more volatile commodities, such as poultry meat.

However, today, integrators use contracts with farmers to outsource the raising of animals for food, shifting the majority of direct costs onto farmers, while retaining strict control over production methods and the distribution of added value throughout the supply chain. For example, integrators often pressure farmers to increase production volumes by taking out loans to modernize their facilities,[22] such as building or enlarging poultry warehouses, or purchasing new ventilation systems to reduce mortality.

Beyond transferring costs and risks, these contracts dictate the ways in which farmers must treat the animals in their care, thus providing the legal mechanism for systemic animal cruelty. Contract terms typically specify detailed instructions on agricultural practices, from manure management to dead animal carcass removal[23] to feeding plans, leaving farmers with

[21] USDA, Economic Research Service. 1996. Farmers' Use of Marketing and Production Contracts, p. 1–2. Available at: https://www.ers.usda.gov/webdocs/publications/40764/18614_aer747a_1_.pdf?v=0.

[22] Sénat. 2021. Suicides en agriculture: mieux prévenir, identifier et accompagner les situations de détresse, Rapport d'information n° 451 (2020–2021). Available at:

[23] MacDonald, J., Perry, J., Ahearn, M., Banker, D., Chambers, W., Dimitri, C., Key, N., Nelson, K., and L. Southard. 2004. Contracts, Markets and Prices: Organizing the Production and Use of Agricultural Commodities. USDA, ERS, Agricultural Economic

little discretion over the care of their animals. In a 2018 French radio program, a calf farmer reported that she was required to have calves drink from a bucket with their heads down, even though it would make the calves sick and bloated as they are used to drinking from their mothers' udders with their heads rising upward.[24] The farmer further added that she felt like a "little soldier" whose sole purpose is "cattle fattening" with no regard for animal welfare.[25] Another farmer filed a lawsuit against a cooperative after the cooperative had mistakenly provided antibiotics, including two prohibited drugs, that caused the death of 25 cows.[26]

What began as a seemingly convenient, mutually beneficial arrangement between growers and processors thus evolved into a mechanism of exploitation—undermining both farmers and animals—by stripping farmers of the autonomy to make management decisions on their own land, including the ability to adopt more humane practices and business models.

2.2 The Legal Distinction Between Contracts With Cooperatives and Those With Integrators

French agricultural law establishes a sharp distinction between contracts concluded with corporate integrators and those concluded with cooperatives.[27] A contract between a farmer and one or several for-profit corporations, explicitly referred to as "integrator," is called an *integration contract*. Under such agreements, the farmer commits "to raising or fattening animals, or producing food of animal origin, as well as to comply with rules regarding livestock management, supply of production means, or sale of

Report No. 837. Available at: https://www.ers.usda.gov/webdocs/publications/aer837/14700_aer837_1_.pdf?v=41061.

[24] France Culture. 2023. Éleveurs enchaînés: "je veux sortir du monde agricole mafieux." Les pieds sur terre, 25 April. Available at: https://www.radiofrance.fr/franceculture/podcasts/les-pieds-sur-terre/modele-integre-eleveurs-enchaines-7292146?fbclid=IwAR3lzhQ7-d7uO4PgwyRadihUOyTuNrCdp9wEWUMLQHYJZ4dO0wHw6X8OzFk.

[25] Ibid.

[26] Ibid.

[27] EU law refers to cooperatives as "producer organizations," even though it fails to provide a harmonized regulatory definition of such producer organizations. Regulation 1308/2013 establishing a common organisation of the markets in agricultural products and repealing Council Regulations 922/72, 234/79, 1037/2001 and 1234/2007, 2013 O.J. (L 347) 671–854.

finished products."[28] Integration contracts are governed by specific regulations designed to protect farmers from abusive practices by corporate integrators.[29]

By contrast, cooperative contracts govern relationships between farmers and agricultural cooperatives.[30] A cooperative contract establishes the membership of a farmer in an agricultural cooperative, which is an organization whose official objective is to provide farmers the ability to "jointly use any methods or means facilitating or developing their economic activity, improving or increasing the results of this activity."[31] Such contracts define both the contribution of the cooperative members to production, and the services provided by the cooperative, even though, in practice, large cooperatives often perform the same functions as corporate integrators. Unlike integration contracts with corporations, cooperative contracts are frequently supplemented by a cooperative's internal rules and bylaws.

Cooperatives and corporate integrators are often presented as pursuing different goals. In 2021, the French High Council of Agricultural Cooperation (*Haut Conseil de la Coopération Agricole*) thus described cooperatives as organizations where "[t]he pursuit of profit is not the ultimate goal. Profits are a lever to develop the cooperative project, strengthen resilience in the face of uncertainties, increase the ability to seize opportunities; in other words, profits ensure the sustainability of the cooperative structure and that of agricultural operations."[32]

However, cooperatives have been operating in ways similar to corporate integrators since legislative reforms beginning in the 1960s, which allowed cooperatives to conduct the same activities as corporate integrators. Specifically, the implementation of the EU's Common Agricultural Policy (CAP)[33] in the 1960s[34] and the 1962[35] French agricultural reforms encouraged farmers to engage in more industrial practices to increase production

[28] Article L. 326–2 of the *Code rural et de la pêche maritime* (Fr.). We translated.

[29] Articles L. 326–1 to L. 326–10 of the *Code rural et de la pêche maritime* (Fr.).

[30] Rules governing relationships between contract growers and cooperatives are codified in Articles L. 521–1 to L. 529–6 of the *Code rural et de la pêche maritime* (Fr.).

[31] Ibid. *Code rural et de la pêche maritime* (Fr.). We translated.

[32] Haut Conseil de la Coopération Agricole. 2021. Le guide des bonnes pratiques – Gouvernance des entreprises coopératives agricoles, p. 17. Available at: https://online.flippingbook.com/view/539241044/ (translation by the authors, free of rights).

[33] See Chap. 3, Nicolas de Sadeleer and Alice Di Concetto, "The Common Agricultural Policy and Animals: Friends or Foes?"

[34] Agricultural Orientation Law No. 60–308 of 5 August 1960.

[35] Law No. 62–933 of 8 August 1962 Complementary to the Agricultural Orientation Law.

volumes, including by facilitating land concentration and a more rational and specialized labor division across the agri-food value chain.[36] These laws also reduced direct state intervention to foster competition between producers. As a result, cooperatives' mission shifted to increasing productivity and competitiveness.[37] Their role ceased to be limited to the collection and marketing of agricultural commodities.[38] Instead, cooperatives actively pursued agronomic progress and increased farming yields.[39]

In the 1970s and 1980s, cooperatives pursued a three-pronged strategy consisting of diversification, concentration, and internationalization.[40] This approach transformed both their purpose and the nature of the contractual relations between cooperatives and their members.[41]

First, between the early 1970s[42] and the 2000s,[43] cooperatives diversified by taking majority stakes in for-profit subsidiaries,[44] enabled by a series of legislative and regulatory reforms.[45] These industrial and commercial activities soon outweighed the cooperative's traditional activities of

[36] Valiorgue, B., Bourlier Bargues, É., and C. Hollandts. 2020. Quelles évolutions de la raison d'être des coopératives agricoles françaises? Regard historique sur un construit social. *RECMA* 358(4): 23–38.

[37] Ibid.

[38] Ibid.

[39] Ibid.

[40] Barraud-Didier, V., Henninger, M.-C., and G. Anzalone. 2012. La distanciation de la relation adhérent-coopérative en France. *Études rurales* 190: 119–130.

[41] Valiorgue, B., Bourlier Bargues, É., and X. Hollandts. 2020. Quelles évolutions de la raison d'être des coopératives agricoles françaises? Regard historique sur un construit social. *RECMA* 358(4): 23–38.

[42] In 1972, cooperatives were authorized to develop activities with the non-cooperative sector up to a limit of 20% of their sales (art. 3 of Law No. 47–1775 of 10 September 1947 on the status of cooperatives and art. L. 522–5 of the *Code rural et de la pêche maritime* (Fr.).). Available at: https://www.legifrance.gouv.fr/codes/article_lc/LEGIARTI000029593648.

[43] Law No. 2001–420 of 15 May 2001 recognizes the possibility for cooperatives to directly or indirectly acquire even majority shareholdings in companies operating business assets (Article L. 124–1 of the *Code de commerce* (Fr.), as amended by Law No. 2001–420 of 15 May 2001 *relative aux nouvelles régulations économiques* (Fr.) (art. 64)).

[44] Valiorgue, B., Bourlier Bargues, É., and X. Hollandts. 2020. Quelles évolutions de la raison d'être des coopératives agricoles françaises? Regard historique sur un construit social. *RECMA* 358(4): 23–38.

[45] Assemblée Nationale. 2022. Rapport d'information sur le secteur coopératif dans le domaine agricole (M. Stéphane Travert et M. Fabien Di Filippo), n° 5040. Available at: https://www.assemblee-nationale.fr/dyn/15/rapports/cion-eco/l15b5040_rapport-information.

assisting farmers' production.[46] As a result, cooperatives began operating in ways as complex as multinational corporations,[47] further blurring the boundaries between agricultural cooperatives and capitalist firms. Second, cooperatives' objectives were also realigned under market concentration dynamics. Multiple mergers of cooperatives produced fewer but larger cooperatives,[48] often affiliated with for-profit corporations through complex corporate structures.[49] These consolidated entities cover large geographic areas and manage large-scale national, and sometimes global, food value chains.[50]

Today, certain large cooperatives function much like multinational corporations, treating contract growers more as workers than as members. Profit distribution is no longer equitable; instead, "the members serve the cooperative."[51] This shift has fueled member concerns about weak representation, opaque financial management, and diminished transparency. For these reasons, it can be challenging to properly distinguish contracts concluded between farmers and corporate integrators, on the one hand, from those concluded with large cooperatives, on the other. Both types of contractual relationships result in restricting farmers' bargaining power and independence, thereby perpetuating the economic dependency of farmers on their trading partner.

From a legal standpoint, the Court of Appeal of Lyon recognized in 1975 that, depending on the context, the contractual relations between farmers and their cooperatives could be no different than the relation between a farmer and corporate integrators.[52] The Court specifically ruled that cooperatives could be qualified as "industrial or commercial compa-

[46] Valiorgue, B., Bourlier Bargues, É., and X. Hollandts. 2020. Quelles évolutions de la raison d'être des coopératives agricoles françaises? Regard historique sur un construit social. *RECMA* 358(4): 23–38.

[47] Assemblée Nationale. 2022. Rapport d'information sur le secteur coopératif dans le domaine agricole (M. Stéphane Travert et M. Fabien Di Filippo), n° 5040, p. 50. Available at: https://www.assemblee-nationale.fr/dyn/15/rapports/cion-eco/l15b5040_rapport-information.

[48] Ibid.

[49] Ibid.

[50] Ibid.

[51] Assemblée Nationale. 2018. Rapport sur le projet de loi, après engagement de la procédure accélérée, pour l'équilibre des relations commerciales dans le secteur agricole et alimentaire et une alimentation saine et durable (n°627), n° 902. Available at: https://www.assemblee-nationale.fr/dyn/15/rapports/cion-eco/l15b0902_rapport-fond.

[52] Cour d'appel de Lyon. Decision of 25 February 1975.

nies" (i.e., corporate integrators) where their operations aligned with those of traditional corporations, noting that "the economic function of cooperatives that transform and sell agricultural products is identical to the one taken on by industrial and commercial businesses."[53] The Court further concluded that, when the legislature exempted cooperatives from certain rules applicable to integration contracts signed with corporations, the legislature did not intend to restrict the effects of the law solely to business relations with corporate entities, but rather to "protect the agricultural producer against businesses with superior economic power."[54]

That reasoning was short-lived. Later, in 1975, the *Cour de Cassation* (Supreme Court) overturned the decision.[55] The French legislature codified this position in 1980,[56] enacting a statute that explicitly excludes relations between agricultural cooperatives and their members from the category of integration contracts, thereby exempting members from the related regulatory safeguards.[57] As a result, under French law, contracts between farmers and cooperatives cannot legally be considered as integration contracts, even though both contract types are the same in practice. This situation tends to shield cooperatives from the need to comply with regulations established to protect contract growers against harmful contractual practices.

3 Harmful Contractual Practices as a Source of Abuse to Farmers and Animals

Agri-food supply chains are riddled with harmful contractual practices that the French legislature has so far failed to remedy, including practices that hinder farmers from transitioning to more humane production methods

[53] Edmond-Noël, M. 1977. Contrat d'intégration/Société coopérative et producteur non adhérent/Absences d'obligation réciproques de fourniture. *Revue Judiciaire de l'Ouest* 1: 84–87 (translation by the authors, free of rights).

[54] Ibid.

[55] Ibid.; Cour de cassation, 1re chambre civile, June 24 1975, *D.* 1976.113, note G. Chesné et E.-N. Martine;

Cour de cassation, 1re chambre civile, November 3 1976, *D.* 1977.521, note G. Chesné et E.-N. Martine;

Cour de cassation, 1re chambre civile, April 27 1978, *JCP* 1979.II.19031, note J. Prévault;

Cour de cassation, 1re chambre civile, May 9 1979, *Bull. civ. I*, n° 139.

[56] Loi n° 80–502 du 4 juillet 1980d'orientation agricole, *Journal officiel de la République française*, 5 July 1980, p. 1734.

[57] Article L. 326–2 of the *Code rural et de la pêche maritime* (Fr.).

or alternative business models. Although contracts with growers typically reference the farmers' obligation to comply with animal welfare laws and regulations, contracts fail to establish the economic conditions necessary to meet such compliance or to support higher standards of animal care. This gap is particularly troubling from an animal welfare perspective, given that current legal frameworks still allow cruel practices.[58] These contracts also often fail to guarantee fair compensation for contract growers (Sect. 3.1), expose them to the risk of unilateral termination by the integrator (Sect. 3.2), and do not ensure that growers have a meaningful ability to terminate the contract on their own initiative (Sect. 3.3).

3.1 Lack of Fair Compensation for Contract Growers

The purchasing prices set by cooperatives and corporate integrators are often below production costs, leaving farmers in precarious economic circumstances. The French legislature attempted to redress this situation by way of the 2021 EGALIM Law.[59] The 2021 EGALIM Law provides that any "upstream" contract must be preceded by a proposal from the agricultural producer, which subsequently forms the basis for negotiations between the parties involved.[60] Furthermore, the law states that the contract must include a price revision clause.[61] In practice, however, the effect of the 2021 EGALIM Law has remained limited. Large agri-food groups retain sufficient market power to dictate non-negotiable prices, effectively neutralizing the law's intended protections.

Farmers working under cooperative contracts also face unfair pay structures. Price setting by cooperatives works in an unusual way, whereby the cooperative essentially sets the prices it will pay to its member farmers in exchange for their production. Specifically, the cooperatives' board of directors sets cooperative members' compensation. Moreover, this board establishes a calculation method for distributing annual surpluses among

[58] The European Institute for Animal Law & Policy. 2022. For a More Humane Union: A Legal Assessment of EU Farm Animal Welfare Laws. Available at: https://animallaweurope.org/wp-content/uploads/For-a-More-Humane-Union.pdf.

[59] Loi n° 2021–1357 du 18 octobre 2021 visant à la protection de la rémunération des agriculteurs (« EGALIM 2 »), *Journal officiel de la République française*, 19 October 2021, as codified in the *Code rural et de la pêche maritime* (Fr.).

[60] Article L. 631–24 of the *Code rural et de la pêche maritime* (Fr.).

[61] Ibid.

members, subject to a vote by the cooperative's general assembly.[62] While this way of operating should, in theory, shield farmers from the risk of their counterparties' abuse of market power, in practice, these boards of directors lack representativeness, and decisions on compensation levels lack transparency due to internal democracy issues.[63] As a result, many members consider compensation levels too low—even though the cooperative model was designed to prevent precisely this problem. Certain multinational cooperatives have parted ways with the original theoretical cooperative model, and this divergence means that members of the cooperative face the same price-setting issues as contract growers working with corporate integrators. Further compounding the problem, both cooperative and integration contracts typically contain exclusivity clauses that prohibit farmers from selling their products to higher-paying buyers, thereby deepening their economic dependency.

Cooperatives also enjoy a significant competitive advantage because they are not subject to corporate income tax.[64] Yet cooperatives rarely use this advantage to benefit their members by purchasing the members' products at a higher price, even though many large cooperatives generate substantial profits.[65]

Farmers in cooperatives are further exposed to underpayment, as a portion of their compensation fluctuates based on performance levels in light of objectives set by their contractual counterpart. These objectives typically do not include animal welfare standards and can even lead to poor animal welfare. For instance, it is common for veal production contracts to include performance bonuses for "white carcasses," incentivizing farmers to feed calves a low-iron diet.[66]

[62] Article L. 521-1-3 of the *Code rural et de la pêche maritime* (Fr.).

[63] Assemblée Nationale. 2022. Rapport d'information sur le secteur coopératif dans le domaine agricole (M. Stéphane Travert et M. Fabien Di Filippo), n° 5040. Available at: https://www.assemblee-nationale.fr/dyn/15/rapports/cion-eco/l15b5040_rapport-information.

[64] Article 207(1), (2), and (3) of the *Code général des impôts* (Fr.).

[65] France Culture. 2018. La fabrique du silence: les agriculteurs sous la pression des industriels. Les Pieds sur terre, 29 January. Available at: https://www.radiofrance.fr/franceculture/podcasts/les-pieds-sur-terre/la-fabrique-du-silence-les-agriculteurs-sous-la-pression-des-industriels-6095588.

[66] The practice is so widespread that EU law prohibits iron-deficient diet in veal production by providing that the food provided to calves "must contain sufficient iron to ensure an average blood haemoglobin level of at least 4,5 mmol/liter." Council Directive 2008/119/EC of 18 December 2008 laying down minimum standards for the protection of calves, 2009 O.J. (L 10), p. 11, Annex I, Point 11.

Lastly, contracts frequently include exclusivity clauses requiring farmers to source all their inputs and services solely from their co-contractor (i.e., the cooperatives or the integrators themselves). These clauses effectively deprive farmers of the ability to negotiate better terms with alternative suppliers. As a result, their production costs—and ultimately their performance—are heavily dependent on the quality of goods (including animals) and services provided by their sole supplier, which simultaneously acts as the buyer of their output. These exclusivity clauses also serve as a powerful means of coercion: integrators can use them to apply financial pressure on farmers who challenge or question contractual practices.[67] Some farmers report receiving low-quality inputs, such as expired or moldy feed and medication, or weak and sickly animals in retaliation for challenging contractual practices.[68] This is particularly problematic, as the co-contractor can then cite poor output performance—caused by these substandard inputs—as grounds to terminate the contract.

3.2 *Discretionary Suspension and Termination*

Integrators and cooperatives are able to unilaterally suspend or terminate their contracts with farmers they consider to be "unproductive." Integration and cooperative contracts usually provide that a company is allowed to suspend a contract with a farmer for unsatisfactory performance, without bonus or compensation. Examples of "poor performance" include a low hen laying rate or low food-to-egg ratio falling 12% below average.[69] Contracts may also be terminated for similar reasons. For example, some clauses permit termination if the feed conversion ratio is too high or if the animals' "average daily growth" is too low. These clauses incentivize farmers to maximize animal size and growth speed while minimizing production costs.

Lastly, integrators and cooperatives often invoke termination clauses abusively. Farmers have reported losing contracts after raising complaints

[67] For a detailed account of similar abusive practices in the US, see Leonard, C. 2014. *The Meat Racket: The Secret Takeover of America's Food Business*. New York: Simon and Schuster.

[68] France Culture. 2018. La fabrique du silence: les agriculteurs sous la pression des industriels. Les Pieds sur terre, 29 January. Available at: https://www.radiofrance.fr/franceculture/podcasts/les-pieds-sur-terre/la-fabrique-du-silence-les-agriculteurs-sous-la-pression-des-industriels-6095588.

[69] This analysis is based on a review of integration and cooperative contracts that are covered by confidentiality agreements.

about their contracting partner or when their production volumes fell short.[70] Integrators and cooperatives thus seem to activate termination clauses to punish farmers with lower production rates or even to silence farmers voicing concerns.

3.3 Contract Growers' Limited Ability to Terminate Contracts

While integrators and cooperatives coerce farmers by threatening contract termination,[71] farmers who want to get out of abusive contractual relationships are often unable to do so due to their significant indebtedness.

Farmers incur significant debt for several reasons. First, integrators and cooperatives strongly encourage farmers to take out long-term loans to finance investments that enable higher production volumes, while farmers often lack adequate information on the implications of such payment plans. Second, banks readily offer long-term loans because farmers typically own their land and use it as collateral in case of default.[72] Banks also place considerable trust in the financial data provided by cooperatives or integrators regarding a farmer's creditworthiness, often without conducting independent verification. Finally, government-backed loan guarantees further encourage banks to view on-farm investment loans as low-risk.

Additionally, farmers under contract with cooperatives often take on short-term loans from their cooperative in the form of "instalment plans," or "cash advances," which must be repaid when leaving the cooperative. Some cooperative contracts even require immediate repayment of all advances if a farmer chooses to leave the cooperative prior to the conclusion of the contractual term. Cooperatives sometimes take farmers' shares in the cooperative as collateral in the event of late payment for feed deliveries.[73] As a result, farmers who fail to repay late fees risk losing their shares in the cooperative.

In addition to being in debt, farmers in cooperatives can face financial penalties under the cooperative's bylaws if they withdraw before the end

[70] France Culture. 2023. Éleveurs enchaînés: "je veux sortir du monde agricole mafieux." Les Pieds sur terre, 25 April. Available at: https://www.radiofrance.fr/franceculture/podcasts/les-pieds-sur-terre/modele-integre-eleveurs-enchaines-7292146.

[71] Legendre, N. 2023. Dans les champs bretons, l'industrie agroalimentaire et la culture de la peur. Le Monde, 4 April.

[72] Legendre, N. 2023. Les paysans bretons dans la spirale du productivisme. Le Monde, 6 April.

[73] Ibid.

of the commitment period.[74] Even when farmers manage to meet these onerous financial terms, many cooperatives and integrators rely on non-competition agreements, committing not to work with farmers who have terminated contracts with competitors.[75] These challenges actively discourage farmers from pursuing more humane production methods.[76]

4 Legislative Responses Failing to End Exploitation on the Farm

Despite abusive contract farming practices being a major factor in the degradation of the treatment of farmers,[77] animals, and the environment,[78] French and EU legislatures primarily approach these issues in fragmented ways. As a result, regulatory measures have largely proved inefficient. Some recent reforms have even aggravated farmers' administrative burden, thereby further undermining their autonomy (Sect. 4.1). At the same time, recent attempts to restore farmers' bargaining power have fallen short of expectations (Sect. 4.2).

4.1 The Aggravation of Farmers' Administrative Burden

Three examples illustrate how state intervention maintains and even worsens farmers' lack of autonomy through over-bureaucratization. First, farmers must navigate an incredibly complex administrative process to obtain the payment of agricultural subsidies (Sect. 4.1.1). Second, the process for obtaining quality production certifications implies numerous screenings and standards that favor farmers under contract whose

[74] Assemblée Nationale. 2022. Rapport d'information sur le secteur coopératif dans le domaine agricole (M. Stéphane Travert et M. Fabien Di Filippo), n° 5040. Available at: https://www.assemblee-nationale.fr/dyn/15/rapports/cion-eco/l15b5040_rapport-information.

[75] France Culture. 2018. La fabrique du silence: les agriculteurs sous la pression des industriels. Les Pieds sur terre, 29 January. Available at: https://www.radiofrance.fr/franceculture/podcasts/les-pieds-sur-terre/la-fabrique-du-silence-les-agriculteurs-sous-la-pression-des-industriels-6095588.

[76] Legendre, N. 2023. Les paysans bretons dans la spirale du productivisme. Le Monde, 6 April.

[77] See Porcher, J. 2002. « Tu fais trop de sentiment », « Bien-être animal », répression de l'affectivité, souffrance des éleveurs. *Travailler* 8(2): 111–134. Available at: https://doi.org/10.3917/trav.008.0111.

[78] See Dumont, B. and P. Dupraz. (eds.) 2016. *Rôles, impacts et services issus des élevages en Europe*. INRA, p. 267–350.

production model is already heavily standardized and fragmented. (Sect. 4.1.2). Third, the rules governing access to crisis management funds are too complex to protect farmers effectively from bankruptcy (Sect. 4.1.3).

4.1.1 A Costly Labyrinthine Process to Obtain Agricultural Subsidies

Filing an application for agricultural subsidies under the Common Agricultural Policy (CAP)[79] involves several challenging steps. In France, farmers must provide a register of their land, identify all land and crop rotations for which they are requesting subsidies, and specify the types of subsidies (among many) they are seeking.[80]

Certain types of subsidies, such as the eco-schemes and Good Agricultural and Environmental Condition of Land, which compensate farmers for good animal welfare and biodiversity practices, require filling out complex administrative forms.[81] The criteria for land eligible for agro-environmental-climate measures and organic payments are equally complex.[82]

This complexity raises farmers' production costs and deepens their dependency on the paid services of accountants, Chambers of Agriculture,[83] and cooperatives, who have the expertise to successfully submit applications for agricultural subsidies on behalf of farmers.[84] The extreme bureaucratization of subsidies thus generates additional costs for farmers, which is particularly concerning since these subsidies exist "to ensure a fair

[79] See Chap. 3, de Sadeleer, N. and Di Concetto, A. "The Common Agricultural Policy and Animals: Friends or Foes?"

[80] Ibid.

[81] Telepac. 2023. Notice de présentation de la télédéclaration du dossier PAC 2023 - Demandes d'aides, éco-régimes et BCAE 8, effectifs animaux et autres obligations. Available at: https://www.telepac.agriculture.gouv.fr/telepac/pdf/tas/2023/Dossier-PAC-2023_telepac_presentation-demande-aides-ecoregime-effectifs-animaux.pdf.

[82] Telepac. 2023. Présentation générale de la télédéclaration du dossier PAC 2023. Available at: https://www.telepac.agriculture.gouv.fr/telepac/pdf/tas/2023/Dossier-PAC-2023_telepac_presentation-generale.pdf.

[83] See dedicated pages of the Ain, Ardèche, or Bretagne *Chambres d'Agriculture* (Fr.).

[84] See the example of Euralis. The reduced access to public services through increased bureaucratization is not specific to the agricultural field. On this topic, see Graeber, D. 2016. The Utopia of Rules: On Technology, Stupidity, and the Secret Joys of Bureaucracy. Brooklyn: Melville House Publishing.

standard of living for the agricultural community, in particular by increasing the individual earnings of persons engaged in agriculture."[85]

4.1.2 *Quality Labels Favoring Contract Farming*

Certain quality food labels, such as the EU organic label,[86] or the French *Label Rouge*,[87] include animal welfare rules.[88] These labels can also increase farmers' income by allowing them to sell their products for a premium price—although such premiums do not always translate into higher incomes for farmers.[89]

However, the process to obtain certifications can be so complex that farmers under integration contracts have a competitive advantage in obtaining them. This is because integration contracts provide strict traceability requirements across the production chain,[90] a system that independent farmers find difficult to replicate. As a result, quality labels such as the organic label and *Label Rouge* risk reinforcing contract farming and its associated unfair trading practices, while independent farmers may find the certification process too onerous.

[85] Consolidated version of the Treaty on the Functioning of the European Union, art 39(1)(b), 2016 O.J. (C 202) 62–63.

[86] EU organic agricultural production standards are regulated under Regulation 2018/848 on Organic Production and Labelling of Organic Products Regulation 2018/848 on organic production and labelling of organic products and repealing Council Regulation (EC) No 834/2007, 2018 O.J. (L 150) 1–92.

[87] The "Label Rouge" standards are governed by French agricultural legislation (Articles L. 641-1 – L. 641-4, and Articles R. 641-1 – D. 641-10-1 of the *Code rural et de la pêche maritime* (Fr.)).

[88] Both labels aim to encourage best practices in agriculture, including animal agriculture. Standards include animal welfare standards, although animal welfare standards feature to a lesser extent in the Label Rouge specifications compared to organic production. For a more detailed analysis of animal welfare in the EU organic regulations, see, e.g., Di Concetto, A. 2023. Farm Animal Welfare and Food Information for EU Consumers: Harmonizing the Regulatory Framework for More Policy Coherence. *European Journal of Risk Regulation* 15(1); Duval, E. 2020. Organic Dairy Cattle: Do European Union Regulations Promote Animal Welfare? *Animals* 10: 1786. https://doi.org/10.3390/ani10101786.

[89] It is particularly the case of sheep in some regions: IDELE. 2016. Viabilité et vivabilité. Références techniques, économiques. Available at: https://idele.fr/detail-article/viabilite-et-vivabilite.

[90] See on traceability issues related to organic certified products: European Court of Auditors. 2019. The Control Systems for Organic Products Has Improved, but Some Challenges Remain. Special Report No. 4/2019, p. 41–44. Available at: https://www.eca.europa.eu/Lists/ECADocuments/SR19_04/SR_organic-food_EN.pdf.

4.1.3 Inadequate Crisis Management Rules: The Example of Depopulation Compensation Funds

Farmers are eligible under national law for payments when their animals are culled ("depopulation"[91]) to prevent the spread of animal diseases.[92] While these compensation systems are designed to protect farmers from bankruptcy, the rules governing these payments often undermine this goal.

When national competent authorities[93] order depopulation campaigns, the French administration[94] responsible for issuing compensation to farmers calculate payments based on the market value of each animal.[95] Farmers are further required to select two experts among a regulatory list, who are subsequently tasked with establishing a joint report that will inform compensation levels.[96] A major limitation of this procedure is that it is lengthy, putting farmers at significant financial risk. French regulations provide an exemption for faster payouts, which allows competent authorities to forgo the expert assessment and immediately issue compensation solely based on the animals' market value.[97] Farmers who face financial challenges tend to prefer this expedited option, even though compensation levels are much lower as the calculation method does not include production costs.

Furthermore, depopulation compensation is granted regardless of production methods, even though some production systems are more disease-prone than others. For instance, highly pathogenic avian influenza spreads more easily on high-density farms that use genetically uniform, less resilient breeds.[98] A more effective way to prevent animal-disease outbreaks

[91] "The process of killing animals for public health, animal health, animal welfare or environmental reasons under the supervision of the competent authority." Council Regulation (EC) No. 1099/2009 of 24 September 2009 on the protection of animals at the time of killing, 2009 O.J. (L 303), p. 8, Article 2(n).

[92] Depopulation is required under Regulation 2016/429 on transmissible animal diseases and Amending and repealing certain acts in the area of animal health as a preventive measure (Articles 7 and 9) or a curative measure when an animal has tested positive for a specific disease (Sect. 4).

[93] In France, the national competent authorities are the *préfectures* (Fr.).

[94] *Direction générale de l'agriculture et de la forêt* (Draaf) (Fr.).

[95] Article 1–1ter of the *Arrêté du 30 mars 2001 fixant les modalités de l'estimation des animaux abattus et des denrées et produits détruits sur ordre de l'administration* (Fr.).

[96] Ibid., Articles 3–4.

[97] Ibid., Article 6bis.

[98] Duval, E. and A. Poinsot. 2024. Le droit de la santé animale: un frein à la transition agroécologique? Étude du dispositif de surveillance et de lutte contre la grippe aviaire en France. *Revue juridique de l'environnement* (in French)

would be to reduce stocking densities,[99] preventing outbreaks and protecting farmers from bankruptcy.

4.2 *Failed Attempts to Restore Balance in Contractual Relations in Animal Agriculture*

In response to farmers' limited bargaining power, the French legislature passed three laws in 2018,[100] 2021,[101] and 2023.[102] The 2018 and 2021 legislative reforms (known as the EGALIM reforms) introduced new measures in French agricultural law, specifically aiming to restore farmers' bargaining power.

The 2018 EGALIM reform sought to ensure a fairer distribution of profits between farmers and integrators,[103] including by changing the rules applicable to contracts between farmers and corporate integrators. The new rules also applied to contracts between farmers and cooperatives if contracts and bylaws produced similar effects as corporate integration contracts, such as providing standards for setting commodity prices.[104] Another major change was the adoption of a set of measures aimed at empowering "producer organizations," a new regulatory category that includes cooperatives.[105] The legislature intended these measures to reinforce the bargaining power of producer organizations in negotiations with food distribution and retail companies.[106]

[99] Ibid. ANSES. 2022. Rapport d'appui scientifique et technique, p. 23–24. Available at: https://www.anses.fr/fr/system/files/LABO2022AST0098Ra.pdf.

[100] Loi n° 2018–938 du 30 octobre 2018 pour l'équilibre des relations commerciales dans le secteur agricole et alimentaire et une alimentation saine, durable et accessible à tous, *Journal officiel de la République française*, 1 November 2018.

[101] Loi n° 2021–1357 du 18 octobre 2021 visant à protéger la rémunération des agriculteurs, *Journal officiel de la République française*, 19 October 2021.

[102] Loi n° 2023–221 du 30 mars 2023 tendant à renforcer l'équilibre dans les relations commerciales entre fournisseurs et distributeurs, *Journal officiel de la République française*, 31 March 2023.

[103] Explanatory statement of Bill No. 627 for Balanced Trading Relations in the Farming and Food Sector, and for a Healthy, Sustainable, and Affordable Diet.

[104] As described in Article L. 631–24, III° and Article L. 631–24, III° of *Code rural et de la pêche maritime* (Fr.).

[105] Article D.554–9 of the *Code rural et de la pêche maritime* (Fr.).

[106] Assemblée nationale. 2018. Rapport, en nouvelle lecture, sur le projet de loi, modifié par le Sénat, pour l'équilibre des relations commerciales dans le secteur agricole et alimentaire et une alimentation saine, durable et accessible à tous (n°1135), n° 1175. Available at: https://www.assemblee-nationale.fr/dyn/15/rapports/cion-eco/l15b1175_rapport-fond.

Because the 2018 reform eventually proved ineffective in addressing farmers' weak bargaining position, the 2021 reform introduced a new set of rules, this time targeting integration contracts. First, because oral contracts were common practice and limited farmers' ability to negotiate, new rules required written and pluriannual sale contracts in the food and farming sectors[107] to allow farmers to understand, verify, and negotiate the criteria used to determine prices.[108] Second, the 2021 EGALIM law mandated the inclusion of a price revision clause in all integration contracts, permitting the parties to freely negotiate the formula and indicators justifying an increase or decrease of the sale price.[109] This new rule aimed to ensure equitable profit distribution in the event of sharp fluctuations in commodity prices. Third, the law required negotiations to be recorded in a written convention listing the parties' obligations and the details of the revision of profit allocation for agricultural commodities.[110] Lastly, the 2021 EGALIM reform essentially prohibited all actors among the agricultural sector from negotiating agricultural raw material prices until the product reaches consumers[111] to avoid reducing the amount ultimately paid to producers of raw material.[112]

Despite these numerous and significant changes in the law, these reforms have brought little change to trade negotiation practices, with a limited impact on the distribution of profits throughout the value chain.

Beyond enforcement gaps,[113] shortcomings also stemmed from substantial limitations in the law. For instance, the 2021 reform required mandatory sale price indexation on raw materials in both farming production contracts and resale contracts, to ensure that farmers would get their

[107] Article L. 631–24 of the *Code rural et de la pêche maritime* (Fr.).

[108] See Delpech, X. 2021. Egalim 2: une nouvelle loi visant à protéger la rémunération des agriculteurs. Dalloz Actualités.

[109] Article L. 631–24, III° of the *Code rural et de la pêche maritime* (Fr.).

[110] Article L. 441–4 of the *Code du commerce* (Fr.).

[111] Article L. 443–8, II° of the *Code du commerce* (Fr.).

[112] DRAAF Occitanie. 2022. Un nouveau cadre pour protéger la rémunération des éleveurs. Available at: https://draaf.occitanie.agriculture.gouv.fr/un-nouveau-cadre-pour-proteger-la-remuneration-des-eleveurs-a6855.html#:~:text=La%20loi%20dite%20. Also see JurisClasseur. 2022. Fasc. 12—Actualité: relations commerciales dans le secteur agricole et alimentaire (Loi n° 2018–938 du 30 octobre 2018; Loi n° 2021–1357 du 18 octobre 2021). JurisClasseur Contrats—Distribution.

[113] Assemblée Nationale. 2022. Rapport d'information sur le secteur coopératif dans le domaine agricole (M. Stéphane Travert et M. Fabien Di Filippo), n° 5040. Available at: https://www.assemblee-nationale.fr/dyn/15/rapports/cion-eco/l15b5040_rapport-information.

fair share of the profits. Yet this requirement had little effect on the outcomes of contractual negotiations, which continued to be dominated by cooperatives and trading partners. The reform fell short of changing these power dynamics by failing to address the issue of contractual exclusivity, leaving farmers obligated to sell their products to a designated cooperative or integrator. Similarly, the reform did not address the challenges farmers face when seeking contract termination with cooperatives or trading partners in order to switch to competitors. The reforms also fell short of regulating the internal governance of cooperatives, which, in their current form, undermine farmers' bargaining power with the cooperative's clients.

5 Conclusion

Advancing the treatment of farmed animals in EU law cannot be achieved in isolation from economic rules structuring the agricultural sector, which shape how power and profits are distributed along value chains. The agricultural crisis that shook the EU between 2022 and 2024 and the subsequent shelving of animal welfare and environmental reforms further illustrate the crucial need to align economic policy with social aspirations.

While legislatures have attempted to redress this dire situation, recent reform efforts have fallen short by adopting a narrow focus on technical fixes. For instance, introducing renegotiation clauses based on cost-calculation methodologies produced little to no effect in the face of deep structural imbalances rooted in longstanding violations of core principles of contract and competition law.

Given the gravity of the situation—within the EU, where an increasing number of governments are promoting anti-liberal agendas, and globally, amid intensifying competition and increasingly aggressive trade relations—both national and EU policymakers must take decisive action. In the specific context of food production, this involves putting an end to the numerous exemptions that cooperatives and integrators have enjoyed until now. Only then will systemic change become possible for the benefit of animals, starting with phasing out the most inhumane practices, which are already widely opposed by farmers, consumers, and citizens alike.

Corporate Social Responsibility and Animal Protection

Clémentine Baldon

1 Introduction

Consumers and investors increasingly expect companies to take environmental, social, and governance (ESG) issues into account when doing business. These expectations cover a wide range of public interest issues, including compliance with human rights, high levels of environmental protection, and the fight against corruption. However, discussions around ESG often fail to take into account one major societal demand: animal welfare, even though improving animal protection is an integral part of corporate social and environmental responsibility (CSR). In the context of expanding and increasingly complex EU legislation on CSR, it remains unclear whether these new rules adequately ensure high levels of animal protection that align with societal expectations.

This chapter provides an overview of existing CSR standards in relation to animal welfare. On the one hand, international organizations have established non-binding animal welfare standards in the context of

C. Baldon (✉)
Baldon Avocats, Paris, France
e-mail: cbaldon@baldon-avocats.com

A.-S. Epstein, A. Di Concetto (eds.), *Animals in EU Economic Law*, The Palgrave Macmillan Animal Ethics Series,
https://doi.org/10.1007/978-3-032-16171-0_6

responsible investment (Sect. 2). On the other, due diligence obligations under French and EU law address animal welfare only minimally, especially when compared to existing non-financial reporting requirements (Sect. 3). For this reason, animal protection advocates have resorted to using law in creative ways, through provisions that indirectly relate to industrial farm animal production, such as environmental liability rules and bans on misleading commercial practices (Sect. 4).

2 Animal Protection and International Responsible Investment Standards

Responsible investment standards consist of numerous principles for companies committed to ethical and sustainable business practices. These principles are formulated by international organizations, such as the United Nations (UN) Principles for Responsible Investment,[1] the UN Global Compact,[2] and the Organisation for Economic Co-operation and Development (OECD)'s Guidelines for Multinational Enterprises.[3] While the majority of these standards focus on the protection of the environment and human rights, some of them have begun to include recommendations on animal protection (Sect. 2.1). Similarly, certain international institutions have recently committed to taking animal welfare considerations into account in their financing activities (Sect. 2.2). In the absence of clear, binding regulations, businesses have introduced voluntary commitments to protect animals, though the effectiveness of these measures remains uncertain (Sect. 2.3).

2.1 A Limited Number of References to Animal Welfare in Responsible Investment Standards

International standards for responsible investment have rarely made explicit references to animal welfare. This situation contrasts with the mounting societal demand for increased levels of animal protection[4] and

[1] United Nations. 2006. United Nations Principles for Responsible Investment.

[2] United Nations Global Compact. 2000. The Ten Principles of the UN Global Compact. Available at: https://unglobalcompact.org/what-is-gc/mission/principles#.

[3] OECD. 2023. OECD Guidelines for Multinational Enterprises on Responsible Business Conduct. Available at: https://doi.org/10.1787/0e8d35b5-fr (in French).

[4] European Commission. 2023. Special Eurobarometer 533 Attitudes of Europeans towards Animal Welfare. Available at: https://europa.eu/eurobarometer/surveys/detail/2996.

the growing recognition of the links between animal protection and sustainable development objectives.[5]

The OECD Guidelines for Multinational Enterprises have attempted to better reflect this societal demand for stronger animal protection considerations. These Guidelines are the authoritative standard for due diligence and the only ones to have an implementation mechanism via National Contact Points (NCPs), which are national offices tasked with receiving and handling complaints against companies, in the event of an alleged breach of the Guidelines. In 2021, the OECD published a review of the Guidelines, assessing their fitness for purpose and identifying opportunities for improvement. In this review, the OECD repeatedly emphasized "the lack of any reference to or recommendations on animal welfare," despite growing public and corporate scrutiny of this issue and the existence of several standards pertaining to animal welfare already in place.[6]

In the draft revision of the Guidelines submitted for public consultation in February 2023, the authors of the Guidelines proposed including animal welfare standards in the chapter on the environment.[7] The drafters of the Guidelines ultimately agreed with the proposal, and, since 2023, the Guidelines have contained the following provision:

[5] See Eurogroup for Animals. 2021. Including Animal Welfare in EU Due Diligence, p. 5. Available at: https://www.eurogroupforanimals.org/files/eurogroupforanimals/2021-12/Including%20Animal%20Welfare%20in%20EU%20Due%20Diligence.pdf. In particular, the "One Health" concept, promoted by the World Health Organization (WHO) and the European Union, recognizes that human and animal health are linked (European Commission. 2017. European Union Action Plan against antimicrobial resistance, p. 4. Available at: https://health.ec.europa.eu/system/files/2020-01/amr_2017_action-plan_0.pdf).

[6] OECD. 2022. Stocktaking Exercise on the OECD Guidelines for Multinational Enterprises, p. 33, 57, and 84. Available at: https://mneguidelines.oecd.org/stocktaking-report-on-the-oecd-guidelines-for-multinational-enterprises.pdf.

[7] "Enterprises should respect domestic laws and regulations pertaining to animal welfare and animal welfare standards that are generally consistent with the World Organisation for Animal Health (WOAH) Terrestrial Code where their activities involve the handling of animals. Good animal welfare requires disease prevention and appropriate veterinary treatment, shelter, management and nutrition, humane handling including transport and humane slaughter or killing." Source: OECD. 2023. Consultation Draft: Targeted update of the OECD Guidelines for Multinational Enterprises and their Implementation Procedures, p. 35, paragraph 82. Available at: https://mneguidelines.oecd.org/consultation-draft-public-consultation-targeted-update-of-the-oecd-guidelines-for-multinational-enterprises.pdf.

> Enterprises should respect animal welfare standards that are aligned with the World Organisation for Animal Health (WOAH) Terrestrial Code. An animal experiences good welfare if the animal is healthy, comfortable, well nourished, safe, is not suffering from unpleasant states such as pain, fear and distress, and is able to express behaviors that are important for its physical and mental state. Good animal welfare requires disease prevention and appropriate veterinary care, shelter, management and nutrition, a stimulating and safe environment, humane handling and humane slaughter or killing. In addition, enterprises should adhere to guidance for the transport of live animals developed by relevant international organizations.[8]

This provision codified into the OECD Guidelines is in fact broader in scope compared to the proposal formulated in the draft, since it refers to animal comfort and the absence of suffering from pain, fear, and distress, which the draft version omitted. The final version is therefore more consistent with a holistic approach to animal welfare. By including animal welfare in the Guidelines, NCPs now have the power to assess complaints related to breaches of animal welfare standards.

Prior to the Guideline revisions of 2023, only certain OECD sectoral guidelines included animal welfare standards. An example of this sectoral inclusion of animal welfare standards would be the OECD–UN Food and Agriculture Organisation (FAO) Guidance for Responsible Agricultural Supply Chains,[9] which urges companies to "strive to ensure that the 'five freedoms' for animal welfare are implemented, i.e. freedom from hunger, thirst and malnutrition, physical and thermal discomfort, pain, injury and disease, fear and distress, and freedom to express normal patterns of behavior."[10] This OECD-FAO Guidance also recommends that companies ensure "high standards of management and stockmanship for animal production [...] in accordance with or exceeding [the WOAH]'s principles."[11] The OECD Due Diligence Guidance for Responsible Supply Chains in the Garment and Footwear Sector furthermore indicates that certain sub-sectors, such as luxury goods, sportswear, and workwear, are

[8] OECD. 2023. OECD Guidelines for Multinational Enterprises on Responsible Business Conduct, p. 38, paragraph 85.

[9] OECD/FAO. 2016. OECD-FAO Guidance for Responsible Agricultural Supply Chains, p. 28. Available at: https://doi.org/10.1787/9789264251052-en

[10] Ibid., p. 29.

[11] Ibid.

faced with animal welfare risks. This Due Diligence Guidance also includes references to the OECD-FAO Guidance.[12]

2.2 *The Role of International Financial Institutions in Promoting Animal Welfare*

In the financial sector, some international public banks take animal welfare into account in their financing strategies. This is the case with the European Bank for Reconstruction and Development (EBRD), as part of its "Environmental and Social Policy," a standard internal to the EBRD that has included animal welfare standards since 2014 and makes reference to national and EU animal welfare regulatory requirements.[13] Furthermore, the EBRD's Environmental and Social Policy explicitly states that the Bank will not finance projects involving, in particular, the "force-feeding of ducks and geese" and "the keeping of animals for the purpose of fur production or any fur production activity."[14]

The World Bank is another public financing institution that has enacted animal welfare standards. Specifically, the World Bank's International Finance Corporation (IFC), whose mission is to develop the private sector, including by investing in companies via loans, equity, debt securities, and guarantees, published a Good Practice Note in 2014 on "Improving Animal Welfare in Livestock Operations."[15] This Note aims to complement the IFC's Performance Standards on Environmental and Social Sustainability, and more specifically the standard on "Conservation of Biodiversity and Sustainable Management of Living Natural Resources."[16]

[12] OECD. 2018. OECD Due Diligence Guidance for Responsible Supply Chains in the Garment and Footwear Sector, p. 48, footnote 14. Available at: https://www.oecd-ilibrary.org/docserver/9789264290587-en.pdf?expires=1723927251&id=id&accname=guest&checksum=84D53E99889295F176CA6A112D46500B.

[13] European Bank for Reconstruction and Development (EBRD). 2019. Social and Environmental Policy, p. 40. Available at: https://www.ebrd.com/news/publications/policies/environmental-and-social-policy-esp.html.

[14] Ibid., p. 10.

[15] IFC. 2014. Good Practice Note: Improving Animal Welfare in Livestock Operations. Available at: https://www.ifc.org/en/insights-reports/2014/publications-gpn-animalwelfare-2014.

[16] IFC. 2012. Performance Standard 6: Biodiversity Conservation and Sustainable Management of Living Natural Resources. Available at: https://www.ifc.org/content/dam/ifc/doc/2010/2012-ifc-performance-standard-6-en.pdf.

An important limitation to both the EBRD's Environmental and Social Policy and the IFC standards is that neither provides for sanctions in the event that the beneficiaries or the banks—the EBRD and the World Bank respectively—breach these policies. While the IFC and the EBRD established complaint mechanisms available to those affected by the financed projects[17] to examine the compliance of the organization's activities with these banks' own internal rules, these mechanisms have not been used extensively. As a result, a number of questions remain unanswered, particularly regarding standing. For example, it is unclear who would the EBRD or IFC consider as "affected" parties in the event of a violation of animal welfare standards, and who would therefore have the right to lodge a complaint. Although these complaint mechanisms do not establish liability and do not grant victims the right to claim compensatory damages,[18] these mechanisms may result in the adoption of corrective measures, such as the temporary suspension or withdrawal of funding.

2.3 Voluntary Corporate Commitments on Animal Welfare

In addition to international organizations' standards, private actors have also signalled a willingness to take animal welfare into account. Companies exposed to significant risks in relation to animals have begun to include animal protection objectives in their CSR policies. An OECD-FAO report on the implementation of the Guidance for Responsible Agricultural Supply Chains, based on a survey of 27 companies, found that 83% of these 27 companies had adopted animal welfare policies.[19]

Corporate communication on non-financial information in the agri-food and luxury sectors further illustrates the increasing importance of

[17] EBRD. 2019. Project Accountability Policy. Available at: https://www.ebrd.com/ipam; IFC/MIGA. 2021. Independent Accountability Mechanism (CAO) Policy. Available at: https://www.cao-ombudsman.org/sites/default/files/documents/CAO%20Policy/ifc-miga-independent-accountability-mechanism-cao-policy.pdf.

[18] Dubin, L. and P. Bodeau-Livinec. 2014. La responsabilité des institutions internationales dans tous ses états. In *Le phénomène institutionnel dans tous ses états: transformation, déformation ou reformation?* Ed. Dubin, L., and M.-C. Runavot, 248–259. Paris: Pedone. See also Richard, V. 2013. Les organisations internationales entre responsibility et accountability: le régime de responsabilité esquissé par la CDI est-il adapté aux organisations internationales? *RBDI* 190(1): 200–201.

[19] FAO. 2019. Pilot project on the implementation of the OECD-FAO Guidance for Responsible Agricultural Supply Chains: Final Report, p. 27. Available at: https://doi.org/10.1787/9789264251052-en.

animal welfare in business. Some companies now include animal welfare risks in their monitoring plans or CSR policies, as part of their due diligence activities. Examples of such companies would be Sodexo,[20] Danone,[21] Coopérative U Enseigne,[22] and Elo (Auchan).[23] Some companies, such as Sodexo,[24] have adopted specific animal welfare policies that apply to their suppliers, while others, like Kering[25] and Danone,[26] have implemented policies across the entire company. These animal welfare policies generally refer to the five fundamental freedoms, as recognized OECD-FAO Guidance for Responsible Agricultural Supply Chain. This is the case for Danone's Animal Welfare Position Paper,[27] Nestlé's Standard for Responsible Purchasing,[28] and Kering's Animal Welfare Standards,[29] which are documents that respectively serve as these companies' animal welfare policies.

Other companies have enacted more specific standards and objectives. For example, Carrefour makes reference to ten "priority objectives" to improve its production methods. These ten objectives are to:

[20] Sodexo. 2022. Universal Registration Document 2022, p. 262. Available at: https://www.sodexo.com/fr/investors/financial-results-and-publications/Universal-Registration-Document.

[21] Danone. 2021. Universal Registration Document 2021, p. 197–201. Available at: https://www.danone.com/content/dam/corp/global/danonecom/investors/fr-all-publications/2021/registrationdocuments/URD2021accessiblefr.pdf#page=199.

[22] Coopérative U. 2022. Coopérative U Enseigne Plan de vigilance de U Enseigne, p. 11–12. Available at: https://www.magasins-u.com/fichiers/pdf_cooperative_u_rapport_rse_2022_plan_vigilance#page=1.

[23] Elo. 2021. Financial Report 2021, p. 33 and 41. Available at: https://groupe-elo.com/app/uploads/2022/03/AUC_2021_RFA_FR_MEL-1.pdf.

[24] Sodexo. 2022. Animal Welfare: Supplier Charter. Available at: https://tracks.sodexonet.com/files/live/sites/com-global/files/02%20PDF/Sodexo_Bien-EtreAnimal_Positionnement.pdf.

[25] Kering. 2019. Animal Welfare Standards. Available at https://www.kering.com/api/download-file/?path=Kering_Animal_Welfare_Standards_2e0e030f40.pdf.

[26] Danone. 2018. Animal Welfare 2016 Position Paper - Updated in March 2018. Available at: https://www.danone.com/content/dam/corp/global/danonecom/about-us-impact/publications/en/2018/2018%2002%2009%20-%20ANIMAL%20WELFARE%20POSITION%20PAPER%202016%20-%20UPDATED%20IN%202018.pdf.

[27] Ibid.

[28] Nestlé. 2018. Nestlé Responsible Sourcing Standards, Section 4.2.10. Available at: https://www.nestle.com/sites/default/files/asset-library/documents/library/documents/suppliers/nestle-responsible-sourcing-standard-fr.pdf.

[29] Kering. 2019. Animal Welfare Standards. Available at: https://www.kering.com/api/download-file/?path=Kering_Animal_Welfare_Standards_2e0e030f40.pdf.

1. eliminate growth-promoting antibiotics and growth hormones
2. eliminate cloning and genetically modified animals
3. "transform cage farming" and limit animal confinement
4. minimize stress during transport and slaughter
5. limit controversial practices and optimize systematic pain management
6. require adequate nutrition
7. require health monitoring
8. eliminate animal testing
9. eliminate animal products sourced from non-farmed animals
10. improve comfort via housing.[30]

While corporate policies demonstrate a growing recognition of animal welfare, "the setting of private standards might amount to making the 'fox guard the hen-house'—which results in low, too low, standards."[31] Furthermore, the monitoring of the implementation of these internal policies is left to the discretion of companies, only a few of which externalize this work to independent private bodies, who monitor compliance mainly through audits. Lastly, because these commitments are voluntary and non-binding, corporations are unlikely to incur liability for violating their own policy, unless the plaintiff identifies a specific violation and can demonstrate it has suffered significant harm as a result of such a violation.

3 Animal Protection in Due Diligence and Sustainability Reporting Rules

In addition to international standards, the EU has enacted legislation that requires companies to exercise due diligence by identifying, monitoring, and mitigating risks throughout their supply chains (due diligence rules); and by reporting on such risks (non-financial or sustainability reporting rules). While the scope of due diligence rules does not cover animal welfare risks (Sect. 3.1), non-financial reporting obligations incorporate

[30] Carrefour. 2021. Garantir le bien-être animal Groupe Carrefour 2021, p. 7–13. Available at: https://www.carrefour.com/sites/default/files/2022-09/Garantir%20le%20bien-%C3%AAtre%20animal_Carrefour%202021.pdf.

[31] Peters, A. 2020. *Animals in international law*. RCADI, vol. 410, p. 187. The Hague: Hague Academy of International Law.

animal welfare. However, these reporting obligations are less binding for companies compared to due diligence rules (Sect. 3.2).

3.1 The Limited Inclusion of Animal Protection in Due Diligence Rules

The French Legislature pioneered the codification of due diligence rules through a 2017 law on the duty of care of parent companies and contractors, called the Corporate Due Diligence Law.[32] This law requires large corporations headquartered and operating in France to implement a monitoring plan aimed at identifying and preventing risks related to human rights, health, personal safety, and environmental impacts arising from their business activities. The French Legislature did not include animal welfare risks in the scope of the 2017 law.

The 2024 EU Directive on Corporate Sustainability Due Diligence (CSDD Directive)[33] largely drew from the 2017 French law. However, unlike the French law, the CSDD Directive does include animal welfare risks. In the preamble of the Directive, the EU Legislature thus states: "Due diligence requirements under this Directive should contribute to [...] creating a toxic-free environment and protecting the health and well-being of people, animals and ecosystems from environment-related risks and negative impacts."[34] The EU Legislature further makes reference to the "One Health" approach, by affirming that "This Directive acknowledges the 'One Health' approach as recognized by the World Health Organization (WHO), an integrated and unifying approach that aims to sustainably balance and optimize the health of people, animals and ecosystems. The 'One Health' approach recognizes that the health of humans, domestic and wild animals, plants, and the wider environment, including ecosystems, are closely interlinked and interdependent."[35]

[32] Law no. 2017–399 of 27 March 2017 on the duty of care of parent companies and ordering companies.

[33] Directive (EU) 2024/1760 on Corporate Sustainability Due Diligence, O.J. L 2024/1760, 7 May 2024.

[34] Ibid., Recital 13.

[35] Ibid., Recital 35. Initially, the Commission's 2022 Proposal for a Directive of the European Parliament and of the Council on Corporate Sustainability Due Diligence and amending Directive (EU) 2019/1937 (COM(2022) 71 final, 23 February 2022) and the position of the Council of the European Union (Proposal for a Directive of the European Parliament and of the Council on Corporate Sustainability Due Diligence and amending

The European Parliament's committees had proposed stronger wording for the CSDD Directive, but the Parliament rejected it at first reading in plenary session. The Committee on the Environment, Public Health and Food Safety, which played a pivotal role in the codification of the reference to the "One Health" concept in the preamble, had suggested including a reference to the five freedoms. The Committee had further proposed including a regulatory definition of the "negative impact on animal welfare" into the Directive, defined as "a negative impact on the welfare of sentient beings resulting from the violation of EU legislation on the protection of animals." Another amendment also proposed broadening the scope of "serious negative impact" to cover a large number of animals, in addition to a large number of people or a large part of the environment.[36] The Legal Affairs Committee presented these proposals for amendments.[37] The European Parliament eventually voted against including these amendments in the Directive. As a result, the Directive only contains indirect references to animals in its preamble, which is not binding. However, these amendment proposals demonstrate the growing importance given to these issues and could further serve as a basis for future legislative initiatives in EU law. Furthermore, the reference to animals' interests in the preamble of the Directive could also produce regulatory effects in the context of litigation efforts seeking to expand the scope of the Directive.

3.2 *The Emergence of Animal Welfare in Corporate Reporting Obligations*

Reporting obligations require companies to communicate about their impacts, thereby forcing them to provide information regarding their practices. Reporting obligations are therefore potentially more powerful than due diligence rules, which merely require corporations to identify and prevent adverse risks. However, the distinction between substantive

Directive (EU) 2019/1937 - General Approach. 6533/22, 30 November 2022) had not included animal welfare to the scope of the Directive.

[36] Opinion of the Committee on the Environment, Public Health and Food Safety for the Committee on Legal Affairs, PE734.465, 10 February 2023.

[37] European Parliament. 2023. Report on the proposal for a directive of the European Parliament and of the Council on corporate sustainability due diligence and amending Directive (EU) 2019/1937, A9–0184/2023, 8 May 2023.

and informational obligations can, at times, be blurred.[38] When it comes to animal welfare, existing legislation on non-financial reporting imposes stricter requirements on corporations compared to due diligence rules. To that extent, non-financial reporting could increase transparency in a way that contributes to improving the treatment of animals.[39]

There are several statutes requiring corporations to report on their impacts on animals, beginning with the laws of Member States. For instance, the 2017 French legislation on the duty of care of parent companies and contractors requires companies to disclose information relating to their societal commitments to animal welfare in their extra-financial performance statements. Specifically, the French law requires that corporations disclose "information relating [...] to [the company's] societal commitments related to [...] animal welfare [...]."[40]

The EU Directive 2022/2464 as Regards Corporate Sustainability Reporting (CSR Directive) expanded the scope of EU non-financial reporting rules to include animal protection among the issues that companies must report on.[41] Before it was revised in 2022, the CSR Directive only required companies to publish the impact of their activities on people and the environment, and contained no reference to animal welfare.[42] The CSR Directive now explicitly references animal welfare, requiring the European Commission to establish reporting standards that companies must comply with when disclosing information regarding their ethics and corporate culture, "including [the areas of] anti-corruption and anti-bribery, the protection of whistleblowers and animal welfare."[43] Although

[38] Epstein, A.-S. 2021. La responsabilité sociale des entreprises, entre obligations de dire et obligations de faire. *Revue des droits de l'homme* 19: 31–36.

[39] Di Concetto, A. and A.-S. Epstein. 2023. EU Consumer Information as a Tool to Regulate the Treatment of Farm Animals. *EU Journal of Consumer Law*: 145–174.

[40] Article L. 225–102-1 III of the French Commercial Code.

[41] Directive (EU) 2022/2464 of the European Parliament and of the Council of 14 December 2022 amending Regulation (EU) No 537/2014, Directive 2004/109/EC, Directive 2006/43/EC and Directive 2013/34/EU, as regards corporate sustainability reporting, 2022 O.J. (L 322) 15–80.

[42] Directive (EU) 2014/95/EU as regards disclosure of non-financial and diversity information by certain large undertakings and groups, 2014 O.J. (L 330) 1–9. Commonly referred to as the "Non Financial Reporting Directive" (NFRD).

[43] Directive (EU) 2022/2464 of the European Parliament and of the Council of 14 December 2022 amending Regulation (EU) No 537/2014, Directive 2004/109/EC, Directive 2006/43/EC and Directive 2013/34/EU, as regards corporate sustainability reporting, art. 1(8) (art. 29b(2)(c)(iii) new), 2022 O.J. (L 322) 15–80.

the European Commission has yet to publish the regulations by way of a delegated act, the CSR Directive encourages corporations to give greater consideration to animal welfare. Furthermore, the inclusion of animal welfare in the scope of information that companies must disclose will likely prompt ESG rating agencies to take this issue into account as well.

While the CSRD Directive represents a significant step towards integrating animal welfare into corporate social responsibility principles, the EU missed an opportunity to advance this integration further when it passed Regulation 2020/852 on the Establishment of a Framework to Stimulate Sustainable Investment (Green Taxonomy Regulation).[44] The Green Taxonomy Regulation aims to direct funding towards sustainable activities, but the EU Legislature did not include animal welfare in its scope. Nevertheless, in June 2023, the European Commission decided to exclude industrial livestock production from the regulatory list of sustainable activities, as enacted by way of a delegated act.[45]

The potential inclusion of industrial livestock production activities on this list had elicited a strong reaction from public interest advocacy groups. In an open letter to the European Commission dated 8 March 2023,[46] a coalition of 25 nonprofit organizations urged the European Commission not to include "industrial livestock farming" in the taxonomy of sustainable activities, using the recommendations from the Platform for Sustainable Finance to bolster their demand.[47] Specifically, the coalition

[44] Regulation (EU) 2020/852 of the European Parliament and of the Council of 18 June 2020 on the establishment of a framework to stimulate sustainable investment and amending Regulation (EU) 2019/2088, 2020 O.J. (L 198) 13–43.

[45] Commission Delegated Regulation (EU) 2023/2486 of 27 June 2023 supplementing Regulation (EU) 2020/852 of the European Parliament and of the Council with technical review criteria for determining under which conditions an economic activity can be considered to contribute substantially to the sustainable use and protection of aquatic and marine resources, the transition to a circular economy, pollution prevention and control, or the protection and restoration of biodiversity and ecosystems, and whether that economic activity does not cause significant harm to any of the other environmental objectives, O.J. L 2023/2486, 21 November 2023.

[46] Letter to Commission on Taxonomy Regulation from 25 civil society organizations, 8 March 2023. Available at: https://www.birdlife.org/wp-content/uploads/2023/03/joint_letter_to_commission_on_forthcoming_taxonomy_delegated_act_on_agriculture_8_mar.pdf.

[47] Platform on sustainable finance: technical working group. 2022. Supplementary: Methodology and Technical Screening Criteria. Available at: https://finance.ec.europa.eu/system/files/2022-11/221128-sustainable-finance-platform-technical-working-group_en.pdf.

insisted that intensive livestock farming is incompatible with climate objectives and pointed out that "producing animal-sourced food (e.g., meat and dairy) emits a larger amount of GHGs than growing crops, especially in intensive, industrial livestock systems" according to the Intergovernmental Panel on Climate Change (IPCC).[48] The coalition further highlighted the many harmful environmental impacts of intensive livestock farming. These include "substantial environmental harms both in the vicinity of the farm and through the upstream activities of producing cereals—wheat, maize, barley and oats—to feed the animals" due to the "abundant use of agro-chemicals, [leading] to soil degradation, biodiversity loss, overuse and pollution of water, as well as air pollution." Additionally, the environmental footprint of imported animal feed is significant, as "[t]he EU relies on the import of huge amounts of soy as animal feed, mostly used in the industrial pig and poultry sectors" and given that "EU soy imports represent the second largest share of EU-driven deforestation which leads to biodiversity loss and the release of high quantities of stored carbon."[49] Although the coalition further argued that "the number of animals raised [should] not exceed the farm's capacity to use their manure to fertilize crops or pasture" for it to be considered sustainable,[50] the exclusion of industrial livestock activities from the regulatory list is provisional, as the explanatory memorandum preceding the Delegated Regulation states that the European Commission will "[continue the evaluation and calibration of the criteria] for certain sectors and activities such as "agriculture, forestry or fisheries [...]."[51] Another potential issue advocates might find challenging to address in the future is the

[48] Letter to Commission on Taxonomy Regulation from 25 civil society organizations includes a reference to the 2022 IPCC Special Report on Climate Change and Land, p. 523.

[49] Ibid.

[50] Ibid.

[51] Draft Commission Delegated Regulation (EU) .../... supplementing Regulation (EU) 2020/852 of the European Parliament and of the Council by establishing the technical screening criteria for determining the conditions under which an economic activity qualifies as contributing substantially to the sustainable use and protection of water and marine resources, to the transition to a circular economy, to pollution prevention and control, or to the protection and restoration of biodiversity and ecosystems and for determining whether that economic activity causes no significant harm to any of the other environmental objectives and amending Delegated Regulation (EU) 2021/2178 as regards specific public disclosures for those economic activities. Available at: https://eur-lex.europa.eu/legal-content/EN/TXT/?uri=pi_com%3AAres%282023%292481554.

lack of a clear definition for industrial livestock production as a production method in EU law.[52]

4 Leveraging Environmental Liability and Unfair Commercial Practices to Advance Animal Interests

In an attempt to overcome the roadblocks imposed by the Legislature on the one hand, and the shortcomings of incorporating animal welfare goals into corporate law on the other, animal advocates have developed novel strategies to hold corporations accountable for animal abuse. In addition to traditional legal actions aimed at holding companies liable for animal cruelty, animal advocacy groups increasingly litigate to find corporations liable for damages their activities might cause to the environment and to animals (Sect. 4.1). Animal advocates also litigate to rid the market of misleading animal welfare claims (Sect. 4.2).

4.1 Environmental Liability

At the international level, two legal actions exemplify the strategic use of environmental protection statutes to advance animal protection goals. These two cases addressed industrial farm animal production activities, with one action utilizing the OECD Guidelines against a company and the other employing the European Convention on Human Rights to challenge the United Kingdom as a State.

In March 2022, a coalition of 17 nonprofit organizations, 26 individuals, and one political party (*Rete Legalità per il Clima*) lodged a complaint with the OECD's National Contact Point (NCP) in Italy against the Cremonini agri-food group.[53] The complainants argued that Cremonini's activities—primarily in beef and pork production and distribution—in Italy have contributed to aggravating climate change, particularly due to the large volumes of methane emitted by farm animals. The complainants

[52] Di Concetto, A. 2024. Les productions animales intensives en droit: état des lieux et perspectives de transition. *Revue juridique de l'environnement* 49: 99–112 (in French).

[53] OECD Watch. 2022. Legalità per il clima on behalf of 45 individuals and Italian CSOs vs. Cremonini S.p.A. Available at: https://www.oecdwatch.org/complaint/legalita-per-il-clima-on-behalf-of-45-individuals-and-italian-csos-vs-cremonini-s-p-a/.

used mediation services by the NCP to obtain Cremonini's disclosure of "information on the climate (and environmental) impacts of its activities, to adjust their risk assessment plans, and to work towards a drastic reduction of methane and other greenhouse gasses and pollutants."[54] An action like this could result in decreasing the number of farm animals if the NCP finds a causal relationship between industrial farm animal production activities and the aggravation of climate change.[55]

The animal protection organization Humane Being, which initiated the "Scrap Factory Farming" campaign, filed a similar application with the European Court of Human Rights against the UK in July 2022.[56] Although the application is not available to the public, Humane Being challenged the UK for failing to regulate factory farming and for failing to take factory farming's climate and public health risks into account. Specifically, the applicants highlighted the risks associated with methane emissions from agricultural sources and the role of soy-based animal feed production in the deforestation of the Amazon. The applicants claimed that the UK government violated Articles 2, 3, and 8 of the European Convention of Human Rights, which respectively guarantee the right to life, the prohibition of inhuman or degrading treatment, and the right to respect for private and family life. However, the Court ruled that Humane Being's legal action was inadmissible in a 2022 ruling.[57]

Because of the 2017 French law on the duty of care of parent companies and contractors, France also provided fertile ground for legal actions aiming to challenge corporations' lack of diligence in limiting the detrimental environmental and climate effects of factory farming. The 2017 French Corporate Due Diligence Law provided the legal basis for a dozen legal actions, two of which challenged animal agriculture activities from a corporate environmental liability perspective.

[54] Ibid.

[55] While this is a positive example of the synergies between environmental protection and animal protection goals, such an alignment is not systematic. For instance, regulating factory farming activities by exclusively taking into account the climate impact of animal agriculture could lead to the development of methanization and biogas, in a way that does not question industrial animal exploitation. See Di Concetto, A., and A.-S. Epstein. 2022. Quelle place pour le bien-être des animaux d'élevage en droit de l'environnement. In *Bien-être et normes environnementales*, ed. Isabelle Michallet, p. 211–225. Paris: Mare et Martin (in French).

[56] European Court of Human Rights. 2024. Climate change. Press Unit, p. 4. Available at: https://www.echr.coe.int/documents/d/echr/FS_Climate_change_ENG.

[57] Ibid.

The first case was brought in 2021 by a coalition of environmental and indigenous groups[58] from Brazil, Colombia, France, and the US against the French supermarket company Casino. The plaintiffs alleged that Casino sourced beef products from South America through production methods that involved deforestation and the expropriation of indigenous lands. The plaintiffs specifically challenged Casino's due diligence plan, arguing that it provided inadequate protection against deforestation risks in its beef supply chain. This lapse, they claimed, enabled Casino to source meat products from JBS, Brazil's largest industrial beef company, despite the company's links to recently deforested areas.[59]

The second case was brought by Brazilian and French nonprofits. *Comissão Pastoral da Terra* (CPT) and Notre Affaire À Tous (NAAT) filed suit against the French bank BNP Paribas in February 2023. The plaintiffs alleged that BNP Paribas was financially supporting Marfrig,[60] Brazil's second largest meatpacking company. The plaintiffs argued that Marfrig was in violation of its commitment not to purchase livestock from farm holdings linked to deforestation in the Amazon after 2009.[61] As a result, they claimed that BNP breached its obligations under the 2017 French Corporate Due Diligence Law[62] by providing financial support to Marfrig. The plaintiffs specifically argued that BNP Paribas' Due Diligence Plan failed to identify the climate risks linked to methane emissions, in particular from cattle farming in Brazil.[63] They also argued that the plan lacked preventive measures to mitigate the environmental damage caused by these activities.

While these two legal actions primarily challenge the environmental impacts of corporations in the context of animal agriculture in Brazil, such litigation could indirectly, but effectively, contribute to higher levels of animal protection. This is because industrial Brazilian meat producers

[58] These organizations are OPIAC, COIAB, FEPIPA, FEPOIMT, Canopée, CPT, Envol Vert, Mighty Earth, Notre Affaire à Tous, France Nature Environnement, and Sherpa.

[59] Radar du devoir de vigilance. Casino en Amazonie. Available at: https://plan-vigilance.org/wp-content/uploads/2021/07/VF-Casino-en-Amazonie.pdf (in French).

[60] Radar du devoir de vigilance. BNP Paribas Déforestation. Available at: https://plan-vigilance.org/wp-content/uploads/2022/12/PDF-BNP-deforestation-FR.pdf (in French).

[61] Assignation devant le Tribunal judiciaire de Paris, 27 Febuary 2023, p. 33. Available at: https://notreaffaireatous.org/wp-content/uploads/2023/02/Assignation-FINALE.pdf.

[62] Ibid., p. 42–45.

[63] Ibid., p. 79.

primarily rely on the use of feedlots—a cruel production method that, in addition to significant environmental and climate risks, raises serious animal welfare concerns.

A similar case was brought in France and could lead to the same indirect, yet positive outcomes for animals. In this case, a nonprofit called *Association Sauvegarde du Trésor-Goëlo-Penthièvre* filed suit against the French government for its failure to regulate the ecological impact of industrial livestock farming. The Rennes administrative Court ordered the French government to compensate for the ecological damage caused by industrial farm producers in the *Baie de Saint-Brieuc.* This ruling came as a result of the failure of the local government to monitor nitrogen emissions, as required by French environmental law.[64] In this ruling, the administrative court further established a correlation between intensive livestock farming and high nitrogen levels, as well as eutrophication, in the Bay of Saint-Brieuc.[65]

Even though the plaintiffs in these legal cases consider animals indirectly—through the protection of the environment and the fight against climate change, rather than considering animals as living beings in their own right—the fight against climate change could become a powerful lever for combating intensive livestock farming, which is also harmful to animal welfare.

4.2 *Consumer Protection Legislation*

Animal and climate advocates who seek to challenge corporations involved in factory farming could use other laws in addition to the 2017 French Corporate Due Diligence Law.[66] One potentially promising lever are rules related to misleading commercial practices.[67] This approach could be useful to the extent that there exist significant procedural uncertainties surrounding the implementation of corporate due diligence legislation.

[64] Tribunal administratif de Rennes, Association Sauvegarde du Trésor-Goëlo-Penthièvre, 18 July 2023, no. 2101565, § 16.

[65] Ibid., § 13.

[66] Barbièri, J.-B., and A. Touzain. 2023. Devoir de vigilance, la porte se referme. *Dalloz actualité*, 13 July 2023 (in French).

[67] In that sense, see also Chap. 10, Isabelle Doussan, "Using Consumer Law to Improve the Treatment of Animals in Industrial Farming Production."

In the EU, commercial speech is governed under Directive 2005/29 Concerning Unfair Business-to-Consumer Commercial Practices in the Internal Market (Unfair Commercial Practices Directive),[68] which prohibits misleading commercial practices.[69] Such practices can include misleading environmental claims ("greenwashing") and misleading animal welfare claims ("humanewashing"). The Unfair Commercial Practices Directive was amended in 2024 by Directive 2024/825 as Regards Empowering Consumers for the Green Transition Through Better Protection Against Unfair Practices and Through Better Information (Environmental Claims Directive)[70] to specifically prohibit misleading environmental claims. In addition to the Unfair Commercial Practices Directive, the European Commission published a proposal for a new Directive on Substantiation and Communication of Explicit Environmental Claims (Green Claims Directive),[71] which would specify the standards that operators must meet when using environmental claims and labels.

Both the Unfair Commercial Practices Directive and the Environmental Claims Directive have the potential to benefit animals directly and indirectly. The Unfair Commercial Practices Directive, as amended by the Environmental Claims Directive, prohibits traders from misleading consumers about the "environmental or social characteristics" of a product,[72] and Recital 3 specifies that the notion of "social characteristics" includes "contributions to social initiatives or ethical commitments, such as animal welfare."[73] The Unfair Commercial Practices Directive may also benefit

[68] Directive 2005/29/EC of the European Parliament and of the Council of 11 May 2005 Concerning Unfair Business-to-Consumer Commercial Practices in the Internal Market, 2005 O.J. (L 149) 22–39.

[69] Ibid., art. 5 to 7, p. 27–29; articles L.121-1 to L.121-4 of the French Consumer Code (Code de la consommation).

[70] Directive (EU) 2024/825 of the European Parliament and of the Council of 28 February 2024 Amending Directives 2005/29/EC and 2011/83/EU to Empower Consumers in the Green Transition Through Better Protection Against Unfair Practices and Better Information, 2024 O.J. L 2024/825.

[71] Proposal for a Directive of the European Parliament and of the Council on the Substantiation and Communication of Explicit Environmental Claims. COM(2023) 166 final, 22 March 2023. The European Parliament and the Council of the EU adopted amendments to the Proposal in March 2024 and June 2024 respectively.

[72] Directive (EU) 2024/825 of the European Parliament and of the Council of 28 February 2024 amending Directives 2005/29/EC and 2011/83/EU to Empower Consumers in the Green Transition Through Better Protection Against Unfair Practices and Better Information, art. 1(2)(a), 2024 O.J. L 2024/826, p. 11.

[73] Ibid., Recital 3, p. 1–2.

animals indirectly by prohibiting the use of private labels that are not based on requirements defined by experts and stakeholders and rely on an independent third party to certify compliance.[74] These new requirements will likely make many misleading animal welfare labels illegal, thus eliminating them from the EU market.

The Proposal for a Green Claims Directive also explicitly refers to animal welfare in the legal act, by requiring companies to "identify whether improving environmental impacts, environmental aspects or environmental performance subject to the claim leads to significant harm in relation to environmental impacts on climate change, resource consumption and circularity, sustainable use and protection of water and marine resources, pollution, biodiversity, animal welfare and ecosystems."[75] These provisions aim to prevent misleading claims about the purported positive effects a practice may have on the environment when there are "trade-offs linked to impacts on biodiversity, ecosystems or animal welfare."[76]

5 Conclusion

There are several ways in which business law can promote higher animal protection levels. At the international level, standards applicable to private operators, such as manufacturers and investors, still lack ambition. At EU and national levels, non-financial reporting rules incorporate references to animal welfare, although there are no due diligence rules that specifically target business practices that affect animal welfare.

This lack of codified rules regarding the consideration businesses must give to animals is not necessarily an obstacle to advocating for more responsible business practices from an animal protection perspective. Businesses that disregard animal welfare expose themselves to reputational damages because of consumers' sensitivity to these issues. These businesses also risk facing legal actions brought on the grounds of environmental liability or unfair commercial practices. The emerging trend in litigation-based advocacy strategies is highlighted by the convergence of objectives pursued by animal and environmental protection advocates.

[74] See Article 2(r) of Directive 2005/29/EC as amended by Directive (EU) 2024/825.

[75] Proposal for a Directive of the European Parliament and of the Council on the substantiation and communication of explicit environmental claims, art. (3)(1)(g), COM(2023) 166 final, 22 March 2023.

[76] Ibid., Recital 19.

Pending the development of international, binding animal welfare standards, advocacy groups are exploring the use of regulatory mechanisms primarily designed for environmental protection as levers against cruel practices, such as industrial farm animal production. Such a strategy can be more effective than traditional litigation tactics that target individual workers or specific corporations. In particular, litigation based on corporate due diligence legislation could lead to the assessment and reform of the entire value chain. Similarly, litigation over misleading commercial practices could result in criminal, civil, and administrative penalties[77] that are often more effective as deterrents compared to criminal penalties under anti-cruelty statutes.[78]

[77] Under French law (Article L.132–2 of the French Consumer Code), the statutory penalty for misleading commercial practices is two years' imprisonment and a 300,000 euro fine. The fine can be increased to 1.5 million euros for corporate persons.

[78] For example, see the Mauléon slaughterhouse case. The former director of the slaughterhouse was sentenced to a six-month suspended prison sentence and the company received a 10,000 euro fine. The sentences for animal cruelty were light. Source: Tribunal correctionnel de Pau, 29 October 2018, no. 16090000004 (in French).

Animals as Consumption Goods

What Can Competition Law Do for Animals? Recent Developments and Future Directions Under the French Competition Authority

Sophie Bresny

1 Introduction

This chapter provides an analysis of how animals are treated under competition law, using the litigation and advisory activities of the French Competition Authority (*Autorité de la concurrence*) as a case study. The French Competition Authority is a government agency tasked with preventing unfair, deceptive, and fraudulent business practices by collecting reports from consumers, conducting investigations, suing companies and people that break the law, developing rules to maintain a fair marketplace, and educating consumers and businesses about their rights and responsibilities. The French Competition Authority is one of the 27 competent national authorities in the EU Member States responsible for enforcing EU competition law.

S. Bresny (✉)
Polynesian Competition Authority, Puna'auia, French Polynesia
e-mail: s.bresny@autorite-concurrence.pf

A.-S. Epstein, A. Di Concetto (eds.), *Animals in EU Economic Law*, The Palgrave Macmillan Animal Ethics Series,
https://doi.org/10.1007/978-3-032-16171-0_7

Following the same logic as in economics, animals in French competition law are primarily considered consumer products, similar to energy or medicines. The approach that the French Competition Authority takes toward animals reflects this view. However, decisions handed down over the past two decades suggest a shift in the way animal welfare is considered in competition law: not merely as a factor in market dynamics, but also as a possible basis for exemptions from standard competition law principles. This evolution reflects how sustainability has gradually become a competitive factor, and that animal welfare is often considered as a facet of sustainability.

This chapter first presents an overview of the French Competition Authority's actions on animal-related issues. It then analyzes how animal welfare has been integrated into the consideration of sustainability in competition law.

2 Animal Welfare in the Decisions and Advisory Opinions of the French Competition Authority

2.1 Data on Cases Involving Animals Litigated Before the French Competition Authority (2001–2023)

Between 2001 and 2023, the French Competition Authority (the Authority) handed down 14 decisions directly or indirectly related to animals, averaging slightly more than one decision every two years. The number of decisions involving animals is thus relatively low, since they represent approximately 1.6% of the total number of decisions on anti-competitive practices handed down by the Authority, which typically issues about 40 decisions per year.

Of these 14 animal or animal-adjacent decisions, three relate to dead animals (marketing of meat or practices in the fishing industry) and do not concern animal welfare issues. Of the remaining 11 decisions, two concern animal insemination (bovine and equine insemination), four are about animal feed (dog and cat food, and vitamins and food supplements for farm animals), two are about animal transport, one is about animal slaughter for meat, and the final two focus on the activities of veterinarians (distribution of medicines and contractual relations between veterinarians and animal welfare organizations in the region of Alsace). These examples illustrate that competition law treats animals and their methods of production (such as slaughter, transport, feed, and reproduction) similarly to any other food product. This approach is consistent with the Authority's mission to ensure the proper functioning of markets by combating cartels and abuses of dominant positions.

Beyond cases involving animals used for human consumption, animals are also treated as consumer products whose prices may be subject to collusion. This is particularly true for companion animals, as the pet market is a dynamic sector valued at nearly €5 billion in the French domestic market alone and approximately €100 billion worldwide, with an estimated annual growth rate of 11%, making it an attractive market for economic operators.

Of the 11 decisions concerning living animals mentioned above, the majority of cases (64%) originated either from the French agency tasked with enforcing consumer and competition law (the Directorate-General for Competition, Consumer Affairs and Fraud Control, DGCCRF) or from a referral by the government (the Ministry of Economy and Finance). Only two cases were referred directly by companies who claimed to have been affected by anti-competitive practices, and two were referred by professional associations or unions. These figures indicate that the DGCCRF is generally best placed to initiate cases involving competition concerns relating to animals, consistent with the DGCCRF's mandate to ensure the proper enforcement of competition law, as well as ensuring food and feed safety for consumers and animals. The DGCCRF also works to ensure compliance with good hygiene practices and investigates chemical or biological agents that may pose risks to humans or animals. As a result, this administration is well positioned to refer cases involving the treatment of animals to the Authority. In the future, however, cases could be initiated by other entities, such as nonprofit organizations, which may be aware of the existence of business practices harming animals and undermining fair competition. This potential development suggests that the Authority has a crucial role to play in training and disseminating knowledge related to competition law.

2.2 *Qualitative Assessment of the Authority's Litigation Activity (2012–2022)*

Between 2012 and 2022, the Authority handed down five decisions that illustrate the treatment of animals in French competition law.

- Decision 12-D-10 of March 20, 2012, concerning practices in the dog and cat food sector

In 2012, the Authority fined three leading companies in the pet food sector (Nestlé, Royal Canin, and Hill's Pet Nutrition) a total of €35 million. The Authority ruled that these companies had restricted competition between 2004 and 2008 in the markets for the sale of dry food (kibble) for dogs and cats sold through specialized distribution channels, including pet shops, garden centers, agricultural supply stores, DIY stores, breeders, and veterinarians.

The Authority specifically found that the companies had engaged in vertical restrictions by colluding with wholesalers to fix resale prices, prohibit "passive sales" (unsolicited sales made spontaneously in response to customer requests), and establish territorial, supply, and customer exclusivity.

In defining the relevant market, the Authority noted that during the period in question (2004–2008), France had approximately 8 million dogs and 10 million cats, and that the three manufacturers accounted for more than 70% of dry pet food sales in specialized retail outlets. The Authority also distinguished between standard products, which are mainly distributed in supermarkets, and high-end products (also known as "premium" or "super-premium"), which are sold exclusively in specialized outlets.

In this decision, the Authority additionally addressed the issue of animal welfare. In Paragraph 4, it notes that "high-end dry foods, referred to as premium or super premium, are intended to facilitate the prevention of certain diseases or to meet physiological needs specific to each stage of [the animal's life]."[1] In Paragraph 260, the Authority further states that "the various agreements implemented by the manufacturers and their wholesalers/distributors concerned products that [...] may give rise to a certain emotional attachment on the part of end consumers, whose brand loyalty tends to make demand less price-elastic. Both wholesalers-distributors and retailers have reported the existence of consumer loyalty." As a result, the agreement between companies had a low impact on consumer demand, meaning that consumers were penalized by paying higher prices for their pet food. As a result, some consumers may have switched to lower-quality pet food sold in supermarkets, while others may have been prevented from accessing higher-quality pet food, potentially affecting the welfare of pets.

[1] All translations in this chapter are ours.

- Decision 13-D-14 of June 11, 2013, on practices implemented in relations between veterinarians and shelters affiliated with the Society for the Protection of Animals (SPA) in Alsace

In 2013, the Competition Authority fined the regional Council of the Order of Veterinarians of Alsace €25,000 and the National Union of Veterinarians in Private Practice in the Bas-Rhin €5000 for establishing an agreement that set the prices veterinarians had to charge animal shelters run by the "Société protectrice des animaux" (SPA) in Strasbourg, and for organizing the sharing of their services with the SPA. As a result of this agreement, participating veterinarians charged identical prices. The price list also covered certain treatments carried out after adoption (particularly spaying and neutering) via vouchers provided by the SPA to adopters. The agreement further stipulated that participating veterinarians would provide their services to the SPA on a rotating basis, according to a schedule established at the beginning of each year. This rotation allowed the participating veterinarians to share the revenue generated by these services equally, and prevented any one veterinarian from capturing the entire market. This agreement particularly affected the SPA in Strasbourg, which was deprived of benefitting from competition for 17 years, between 1991 and 2008, as almost all veterinarians in the Strasbourg metropolitan area had signed the agreement.

In its decision, the Competition Authority takes animal welfare into account in several paragraphs (147–150) when assessing whether the Council could qualify for an exemption from the rules on cartels and abuse of dominant position. The Council had argued the agreement ensured positive economic outcomes, including through the creation or maintenance of jobs, and that it had reserved a fair share of the resulting profit for users, without substantially eliminating competition for the services provided.

The Council specifically argued that the agreement had enabled the proper care of animals for the lowest possible costs. However, the Authority pointed out that the Council failed to demonstrate why the agreement, as opposed to free competition between veterinarians, would have achieved such a positive outcome. The Authority also notes that the managers of other animal shelters in the Alsace region have expressed satisfaction with the quality of the services provided by veterinarians charging lower rates than those applied to the Strasbourg SPA, and that none had ever indicated any need for a price-unifying agreement. In Paragraph 170, the

Authority thus concludes that "the facts are all the more serious given that the persons affected or likely to be affected by the practices, namely mainly the Strasbourg SPA and other SPAs in the Alsace region, are nonprofit, public interest organizations and derive their resources mainly from individual donations."

The Authority therefore adopted a pragmatic approach to the case, taking into account the specific nature of the entity harmed by the practices, namely animal shelters. This decision is particularly important for animal welfare, as the potential price increases resulting from the price list could have had an impact on the overall volume of care that the Strasbourg SPA was able to provide.

- Decision 18-D-15 of July 26, 2018, on practices in the veterinary medicine distribution sector

In a 2018 decision, the Authority fined wholesale distributors of veterinary medicines and their federation (*Fédération de la Distribution du Médicament Vétérinaire*, FDMV) €16 million for implementing a customer sharing/non-aggression pact and an agreement on vaccine delivery prices and invoicing for veterinarians.

The decision notes that France is one of the EU's leading livestock producers and has the largest population of companion animals in Europe, making France the leading European market for veterinary medicines. Additionally, wholesale distributors account for 90% of the distribution of veterinary medicines to veterinarians, livestock farmers' groups, and pharmacists.

With this decision, the Authority concludes that all distributors and the FDMV colluded to take advantage of the animal health crisis caused by the rapid spread of bluetongue between 2007 and 2010, which led to the implementation of health emergency measures, including three vaccination campaigns in France between spring 2008 and autumn 2010.

Taking advantage of the epidemic context, all wholesale distributors and the FDMV agreed on the prices they would submit to public authorities to maximize payments for the logistical services involved in distributing vaccines. This strategy enabled the companies to earn substantial illegal gains. According to the distributors' own statements, they received 4 cents per vaccine distributed, which is significantly higher than the actual costs, which ranged from 1 cent to 2.5 cents. Furthermore, the distributors and the FDMV jointly set the prices charged to veterinarians during the second and third compulsory vaccination campaigns.

The Authority notes that the FDMV played a pivotal role in planning and organizing the strategy. During the first vaccination campaign, the FDMV participated in developing and disseminating the calculation method that enabled wholesale distributors to overestimate their costs in order to obtain higher payments. During the two subsequent campaigns, the FDMV further used its influence to convince distributors not to agree to a reduction in payments claimed from the administration compared to those obtained during the first campaign.

However, in its decision, the Authority does not directly address the issue of animal welfare. In Paragraph 206, the Authority notes that "the distributors took advantage of the health emergency linked to the rapid spread of bluetongue and the absence of a call for tenders to deliberately mislead the public purchaser about the true logistical costs of delivering the vaccines incurred by each of them, thereby compromising the proper use of public funds," thus primarily taking into account the economic interests of the public administration and farmers. Yet, the inflated vaccination costs resulting from the agreement also had animal welfare consequences, as bluetongue causes significant suffering in sheep, goats, and cattle. Affected animals are unable to feed properly and develop pulmonary edema. In addition, young animals can be born with malformations. As a result of this agreement between distributors and the resulting inflated costs, some farmers chose not to vaccinate their animals, potentially impacting their welfare.

- Decision 19-D-03 of January 16, 2019, on practices implemented in the cross-Channel transport of day-old chicks

In 2019, the Competition Authority issued a decision following a referral from two breeders of day-old chicks intended for use in hunting. The referral arose after cross-Channel shipping companies refused to transport the chicks due to a campaign by anti-hunting organizations, which exposed the companies to the risk of a boycott.

The chick market is significant in France, as French farms that breed animals for hunting purposes produce approximately 20 million birds for the domestic market. These producers also export 10 million eggs and day-old chicks, mainly to England and Spain, in addition to large mammals used in canned hunting.

In this case, two chick farmers filed a complaint with the Authority, requesting injunctive measures to address the breach of shipping contracts by cross-Channel shipping companies, which they also accused of abusing a dominant position.

The Authority did not rule on the merits of the case due to a lack of evidence, particularly with regard to the definition of the relevant market. Specifically, the Authority did not rule on the question of the substitutability between modes of transport (air, rail, or sea). Had the Authority examined this issue, it might have taken into account the welfare and health of the animals, as suggested by Paragraph 34 of the decision, where the farmers argued that "the maritime transport service is objectively necessary to transport the chicks at the best price and under the best sanitary conditions."

Had animal welfare been given greater consideration, the Authority would have likely found that the shipping companies abused their dominant position, on the grounds that sea transport is the least cruel mode of transport for day-old chicks and that refusing to transport the animals by sea would result in longer transport times. However, it is also possible that the additional costs of arranging an alternative transportation mode could have halted the shipment of chicks entirely, which would have been even more beneficial to their welfare.

- Decision 22-D-05 of February 15, 2022, on practices implemented in the live animal transport sector by air freight

In a 2022 decision, the Authority fined Goldenway International Pets (GIP) €65,000 for abuse of a dominant position. This decision follows a report submitted by the DGCCRF and an investigation conducted with the assistance of the Polynesian Competition Authority into the air freight transport of companion animals from mainland France to Polynesia.

In this case, people wishing to travel or relocate to French Polynesia with their companion animals were required to have their animals' health checked by an official veterinarian, a procedure carried out during the animal's quarantine period. Because French Polynesia does not have the infrastructure to carry out this check on its territory, the animals had to remain in the only metropolitan quarantine station authorized by the Polynesian authorities, which was built and operated by the company GIP on the mainland, in Normandy. Travelers rarely used alternative quarantine facilities, such as quarantine in a territory through which the animal would be transiting (New Caledonia, New Zealand, Australia, or Hawaii), due to the complexity of the administrative formalities, higher costs, and animal welfare risks.

As a result, anyone traveling with a companion animal to French Polynesia from mainland France was required to have the animal quarantined at the GIP station. GIP exploited this requirement to force the sale of a package deal, which included road transport services from the quarantine station to Roissy-Charles de Gaulle airport and the animal's plane ticket from France to Polynesia. Companion animal owners therefore had no choice but to use GIP for all three services. Despite repeated requests from consumers wishing to purchase the transport and air freight services for their animals separately, and despite offers from other providers at lower rates, GIP consistently refused to unbundle the services. By linking its de facto monopoly on quarantine with road and air transport, GIP prevented any competition to the detriment of both competitors and consumers.

The Authority refers to animal welfare five times in its decision (paragraphs 80, 174, 269, 361, and 415). First, the Authority rejects the alternatives to quarantine proposed by GIP in New Caledonia, New Zealand, Hawaii, California, and Japan on the grounds that "These alternatives are effectively ruled out by consumers because they are, in particular, very costly, administratively burdensome, logistically and sometimes linguistically complex, and more dangerous for the health and welfare of the animals."[2] The Authority also points out that these alternatives significantly increase transport times and are not acceptable to individuals.[3] Paragraph 269 also explicitly refers to animal health. Lastly, the decision refers to the danger of alternative solutions to animal health and welfare by specifying that "due to the strong emotional bond between them and their animals, these customers had no other solution for importing them than to accept the conditions and prices imposed by GIP."[4]

The Authority further concludes that the abuse of a dominant position, along with the costs borne by consumers, led to the situation where "other consumers have had to give up, at least temporarily, taking their pets to Polynesia." The Authority uses this finding as a basis to qualify GIP's dominant position and to consider the infringement to be particularly

[2] Paragraph 361.

[3] Paragraph 74. Note that the Authority refers to "individuals" and not consumers here.

[4] Paragraph 415.

serious. However, the Authority does not consider animal welfare as a stand-alone factor in its decision. Rather, animal welfare is linked to consumer welfare, as consumers suffer from higher costs and a more complicated procedure.

This decision is likely to have positive effects on both competition and the welfare of companion animals, as it will reduce the cost of transporting them, thereby helping prevent abandonment due to owners' limited financial means—at the time of the decision, transporting an animal cost twice as much as an adult airline ticket.

2.3 Overview of the Competition Authority's Other Missions in the Field of Animal Welfare

In conducting its mission to enforce competition law and maintain economic public order, the Authority has several tools beyond sanctioning cartels and abuses of dominant positions. First, the Authority issues opinions, either on referral or on its own initiative, on draft legislation, proposed reforms, or any competition issue relevant to public debate. These opinions may include recommendations aimed at improving the functioning of competitive markets (2.3.1). Second, the Authority monitors mergers, acquisitions, and the creation of joint ventures exceeding a certain size prior to their implementation. Following its review, the Authority may grant authorization, with or without conditions, or impose a veto (2.3.2).

2.3.1 Animal Welfare in the Advisory Activities of the Competition Authority

The Authority has not issued any opinion specifically concerning animal welfare. However, the Authority discussed animal welfare in a 1998 opinion relating to a request from the National Union of Dog Professionals (*Syndicat national des professionnels du chien*) on various competition issues encountered by its members from nonprofit organizations in the management of stray dogs.[5]

The professional union argued that companies, associations, or societies operating in the same market must be subject to the same conditions of competition. The union considered that animal welfare nonprofit organizations were distorting competition in the animal shelter management

[5] Opinion No. 98-A-02.

market because of their tax-exempt status, which allowed them to offer lower prices for services of capturing, transporting, and sheltering dogs.

The Authority emphasizes that animal welfare nonprofits carry out general interest work, which in practice they are virtually the only ones providing. Given their social purpose, the nonprofits typically build and manage shelters but rarely capture the animals. By contrast, for-profit companies involved in the construction and management of shelters typically provide integrated services that include both the capture and transport of animals. The Authority therefore concluded that the differing legal statuses of entities operating in this sector, reflecting their different activities and missions, are not likely to distort competition in the market for stray management services, where both animal protection nonprofits and commercial companies are active.

2.3.2 Animal Welfare in Merger Control

The Authority issued two merger control decisions in 2011 and 2021 that expressly refer to animal welfare. In these cases, considerations of animal welfare were incorporated into the delineation of the relevant market, thereby narrowing down its geographical scope. These two decisions specifically referred to the welfare of animals intended for slaughter during transport.

In the 2011 decision,[6] the Authority took animal welfare into account in defining the geographic market for the collection of live animals for slaughter, which was found to be "most often local or regional in scope, resulting from regulatory constraints relating to animal welfare and the significance of transport costs."[7] The 2021 decision[8] took animal welfare into account in defining the geographic market for the collection of live

[6] Centrale Coopérative Agricole Bretonne. 2011. Décision n° 11-DCC-154 du 24 octobre 2011 relative à la prise de contrôle exclusif de la société Financière du Forest, holding du groupe GAD, par la société Centrale Coopérative Agricole Bretonne. Available online: https://www.autoritedelaconcurrence.fr/sites/default/files/commitments/11DCC154decision_version_publication.pdf.

[7] Ibid., paragraph 30.

[8] Autorité de la concurrence. 2021. Décision n° 21-DCC-65 du 14 avril 2021 relative à l'acquisition par la société LDC Volaille du groupe Ronsard. Available online: https://www.autoritedelaconcurrence.fr/sites/default/files/integral_texts/2021-05/20-085_decision_21-dcc-65-publique.pdf.

poultry for slaughter, ultimately finding the market to be local rather than national in scope.[9]

3 The Integration of Animal Welfare in Competition Law: Prospective Changes

Recent EU competition legislation in favor of animal welfare (3.1.) and the broader significance of sustainability considerations, of which animal welfare is a component (3.2), both suggest that the Authority will increasingly assess and protect animal welfare and health in the years ahead.

3.1 Recent EU Competition Legislation in Support of Animal Welfare Objectives

Three recent EU competition law instruments are likely to further embed animal protection objectives into competition law.

3.1.1 The European Commission's Guidelines on Horizontal Co-Operation Agreements

A fundamental principle of constitutional EU law ("primary law") provided in Article 101(1) of the Treaty on the Functioning of the EU is a prohibition on "all agreements between undertakings, decisions by associations of undertakings and concerted practices which may affect trade between Member States and which have as their object or effect the prevention, restriction or distortion of competition within the internal market."[10] However, Article 101(3) provides an exemption to this prohibition for restrictive agreements that produce objective economic benefits outweighing the negative effects of restricted competition. The EU legislature has codified such an exemption in secondary legislation, either on a case-by-case basis or for categories of agreements and concerted practices.

In 2023, the European Commission adopted guidelines listing four cumulative conditions that economic operators must satisfy in order to benefit from this exemption:

[9] Ibid., paragraph 11.

[10] Consolidated Version of the Treaty on the Functioning of the European Union, art 101, May 9, 2008 O.J. (C 115) 88–89.

1. Their agreements or practices must contribute to improving the production or distribution of products, or to promoting technical or economic progress;
2. The restrictions must be indispensable to the attainment of these objectives;
3. Consumers must receive a fair share of the resulting benefits;
4. The agreement must not afford the parties the possibility of eliminating competition in respect of a substantial part of the products concerned.

These guidelines further provide guidance on how the European Commission is likely to interpret these conditions.[11] Specifically, when assessing sustainability agreements, these guidelines provide that competitors who "may wish to agree on certain standards to improve animal welfare (e.g. standards to provide animals with more space and better living conditions)" are allowed to "adopt and comply with certain sustainability standards."[12]

3.1.2 The 2022 Guidelines on Vertical Agreements

In 2022, the Commission followed a similar approach regarding vertical agreements through Regulation 2022/720 on the Application of Article 101(3) of the Treaty on the Functioning of the European Union to Categories of Vertical Agreements and Concerted Practices.[13] A vertical agreement is a practice implemented by manufacturers or suppliers concerning the resale of their products. Common practices in this area include price fixing, exclusive distribution, and territorial or other geographical restrictions. Under an exclusive distribution or territorial exclusivity system, the manufacturer authorizes only one distributor to market the product. In assessing the legality of such practices, competent competition authorities must determine whether this exclusivity confers market power on the distributor by limiting inter-brand competition.

[11] European Commission. 2023. Communication Guidelines on the applicability of Article 101 of the Treaty on the Functioning of the European Union to horizontal cooperation agreements, C/2023/4752 final.

[12] Ibid., paragraph 538.

[13] Commission Regulation (EU) 2022/720 of 10 May 2022 on the application of Article 101(3) of the Treaty on the Functioning of the European Union to categories of vertical agreements and concerted practices, O.J. (L 134) 4–13.

Regulation 2022/720 and its accompanying guidelines[14] thus introduced considerations relating to the ways in which competent authorities should consider sustainability objectives in the assessment of vertical agreements. These guidelines are particularly relevant to animal welfare, as the Commission defines sustainability as a "notion" encompassing "addressing climate change, limiting the use of natural resources, reducing waste, and promoting animal welfare."[15]

3.1.3 Regulation 1308/2013 on the Common Organization of Agricultural Product Markets

In 2021, the EU legislature amended Regulation 1308/2013 Establishing a Common Organisation of the Markets in Agricultural Products[16] to introduce a new derogation from EU competition rules. The Regulation provides that "Article 101(1) TFEU shall not apply to agreements, decisions and concerted practices of producers of agricultural products that relate to the production of or trade in agricultural products and that aim to apply a sustainability standard higher than mandated by Union or national law, provided that those agreements, decisions and concerted practices only impose restrictions of competition that are indispensable to the attainment of that standard"[17]—further specifying that "sustainability standard" means "a standard which aims to contribute to one or more of the following objectives [...] animal health and animal welfare."[18]

3.2 Animal Welfare as an Element of Sustainable Development

In 2020, the Authority designated sustainable development as one of its enforcement priorities. As a result, the Authority advises companies seeking guidance on this subject, for example when they are considering entering a sustainability agreement. The Authority further contributes to initiatives on this topic at EU level, particularly in the context of the

[14] Guidelines on Vertical Restraints 2022/C 248/01, 2022 O.J. (C 248) 1–85.

[15] Ibid., Recital 8.

[16] Regulation 1308/2013 on the common organization of the markets in agricultural products and repealing Regulations 922/72, 234/79, 1037/2001 and 1234/2007, 2013 O.J. (L 347) 671–854. See Chap. 3, Nicolas de Sadeleer and Alice Di Concetto, "The Common Agricultural Policy and Animals: Friends or Foes?"

[17] Regulation 1308/2013 on the common organization of the markets in agricultural products, art 210a(1), 2013 O.J. (L 347) 109 (consolidated text).

[18] Ibid., art 210a(3)(c).

European Green Deal. The creation of a Head of the Sustainable Development Network at the Authority in 2022 confirmed the significance of these efforts. A culture of sustainable development is thus spreading across the organization and through the work of the 20 rapporteurs, and one of the Authority's objectives is to encourage nonprofit organizations to refer matters to the Authority as a way to promote knowledge and enforcement of competition law and policy.

4 Conclusion

National competition authorities have a key role to play in improving the treatment of animals in economic law. However, this role depends in part on a change in consumer behavior, as competition authorities can consider animal welfare in their competition analysis only to the extent that it constitutes an essential competitive factor for consumers—similar to the way sustainability is increasingly treated by both consumers and competent authorities. In practice, this means that authorities are likely to adjust their approach to align consumer welfare with animal welfare only if consumers recognize that two given product offers are not interchangeable due to the negative impact that one of these products might have on animals.

Animal Protection in Public Procurement Rules

Anne-Laure Meynckens

1 Introduction

Public procurement refers to the purchase of goods and services from external, usually private, sources by governments or public agencies. Public procurement by public administrations typically involves a bidding process, and procurement agreements are always contractual.

Despite its complexity, the topic of the inclusion of animal welfare standards in public procurement rules represents one of the most promising action levers for advancing animal protection. Such standards can raise awareness among producers, buyers, and consumers about animal ethics, while also encouraging more sustainable and humane production and consumption practices. This is particularly significant in France, as in most EU countries, where the government purchases large volumes of goods

A.-L. Meynckens (✉)
Agence Drôle de Zèbre, Paris, France

Agence Symbiocène, Paris, France

A.-S. Epstein, A. Di Concetto (eds.), *Animals in EU Economic Law*, The Palgrave Macmillan Animal Ethics Series,
https://doi.org/10.1007/978-3-032-16171-0_8

and services. By requiring vendors to comply with specific standards, governments can directly influence production methods.[1]

The inclusion of animal welfare standards in public procurement rules can be mainstreamed at all levels of governance but is especially effective at local level. Under EU law, animal welfare has recently gained recognition as a relevant selection standard in public procurement. Specifically, Directive 2014/24 on public procurement[2] explicitly refers to the use of animal welfare standards as a means to promote animal welfare measures.[3] National legislatures have codified EU legislation accordingly and now require public authorities to source their supplies from producers who comply with quality standards—including animal welfare standards—that go beyond minimum legal requirements. France incorporated Directive 2014/24 into national law in 2018.[4]

The French approach also reflects societal expectations, which align with the creation of animal welfare officers at the local level. Since the 2020 municipal elections in France, new animal protection mandates and policies (such as animal welfare, the treatment of animals, animals in urban settings) were created within municipal councils, responding to citizens' concerns. A 2020 survey[5] found that 72% of French citizens believed it was important for candidates in the 2020 municipal elections to commit to animal protection measures in their communities. The same percentage believed it was important for mayors of municipalities to appoint deputies responsible for animal protection.

This chapter presents a case study on the inclusion of animal welfare standards in public procurement policies in France, based on empirical but non-exhaustive examples identified during research conducted in 2022.

[1] Directive 2014/24/EU on public procurement and repealing Directive 2004/18/EC, 2014 O.J. (L 94) 65–242. Also see Coller Animal Law Forum. Public procurement Factsheet. Available at: https://calf.law/factsheets/public-procurement. Accessed 29 April 2025.

[2] Directive 2014/24/EU of the European Parliament and of the Council of 26 February 2014 on public procurement and repealing Directive 2004/18/EC, 2014 O.J. (L 94) 65–242.

[3] Ibid., Recital 98.

[4] Loi n° 2018–938 du 30 octobre 2018 pour l'équilibre des relations commerciales dans le secteur agricole et alimentaire et une alimentation saine, durable et accessible à tous, *Journal officiel de la République française*, November 1, 2018.

[5] IFOP, L214. 2019. Les Français et la prise en compte de la cause animale dans les politiques municipales. Available at: https://www.politique-animaux.fr/fichiers/sondage-animaux-elections-municipales-2020.pdf.

2 Legislative and Regulatory Framework

Public procurement in France is governed by both EU and national law (2.1), which together provide a broad scope for the implementation of animal welfare rules in public contracts (2.2).

2.1 Relevant EU and French Legislation

The inclusion of animal welfare standards in public procurement is consistent with the objectives set out in EU legislation, starting with Article 13 of the Treaty on the Functioning of the European Union, which states:

> In formulating and implementing the Union's agriculture, fisheries, transport, internal market, research and technological development and space policies, the Union and the Member States shall, since animals are sentient beings, pay full regard to the welfare requirements of animals, while respecting the legislative or administrative provisions and customs of the Member States relating in particular to religious rites, cultural traditions and regional heritage.

This general provision is partially but directly implemented in EU secondary law, through directives and regulations pertaining to animal welfare, such as Directive 2010/63/EU on the Protection of Animals Used for Scientific Purposes, which refers to the intrinsic value of animals in recital 12.[6] Indirect implementation occurs through legislation such as Directive 2014/24/EU on Public Procurement[7] and Directive 2014/25/EU on Procurement.[8] Directive 2014/24, which reforms public procurement rules in the EU, states in recital 98: "Contract performance conditions might also be intended to favour the implementation of measures for [...] the protection of the environment or animal welfare [...]."

[6] Directive 2010/63/EU of the European Parliament and of the Council of 22 September 2010 on the protection of animals used for scientific purposes, Recital 12, 2010 O.J. (L 276), p. 34.

[7] Directive 2014/24/EU of the European Parliament and of the Council of February 26, 2014 on public procurement and repealing Directive 2004/18/EC, 2014 O.J. (L 94) 65–242.

[8] Directive 2014/25/EU of the European Parliament and of the Council of February 26, 2014 on procurement by entities operating in the water, energy, transport, and postal services sectors and repealing Directive 2004/17/EC, 2014 O.J. (L 94) 243–374.

This recital was transposed in French law via Article 62 of a 2016 Decree,[9] subsequently codified in 2018 in Article R2152-7 of the Public Procurement Code,[10] which states:

> In order to award the contract to the bidder or bidders who have submitted the most economically advantageous tender, the purchaser shall base its decision:
>
> 1° Either on a single criterion, which may be: a) The price, provided that the sole purpose of the contract is the purchase of standardized services or supplies whose quality is not subject to variation from one economic operator to another;
>
> b) Cost, determined according to a comprehensive approach that may be based on the life-cycle cost defined in Article R. 2152-9;
>
> 2° Either on a number of non-discriminatory criteria related to the subject matter of the contract or its performance conditions, including the criterion of price or cost and one or more other criteria including qualitative, environmental, or social aspects. These may include the following criteria: a) Quality, including […], animal welfare (…).[11]

Although this provision does not require public administrations to take into account animal welfare when awarding contracts to vendors, it expressly permits them to do so. This reflects the principle that "legislation should provide guidance, not just be a mechanism for punishment."[12]

2.2 *Scope of Animal Welfare Rules in Public Procurement Rules*

2.2.1 *Types of Contracts*

Public procurement contracts—representing 8% of France's GDP in 2024[13]—are tightly regulated, primarily under the Public Procurement Code. Article R. 2152-7 of the Public Procurement Code applies to public procurement contracts, which are "commercial contracts concluded by

[9] *Décret n°2016–360 du 25 mars 2016 relatif aux marchés publics* (Fr.), Article 62.

[10] *Code de la Commande Publique* (Fr.), Article R2152-7.

[11] We translated.

[12] Radford, M. 2001. *Animal Welfare Law in Britain: Regulation and Responsibility.* Oxford: Oxford University Press.

[13] Intercommunalités de France and Banque des Territoires. 2024. Public procurement barometer – Results for 2023 and H1 2024. Available at: https://www.intercommunalites.fr/publications/barometre-de-la-commande-publique-resultats-de-2023-et-du-s1-2024/.

a purchaser or an authority to meet its needs in terms of works, supplies, or services, with one or more economic operators."[14] Public procurement contracts can therefore be very diverse, being either public contracts, defined as contracts "concluded by one or more purchasers subject to the [Public Procurement Code] with one or more economic operators, to meet their needs for works, supplies or services, in return for a price or any equivalent,"[15] or concessions, which refer to contracts "whereby one or more authorities subject to the [Public Procurement Code] entrust the execution of works or the management of a service to one or more economic operators, to whom a risk related to the operation of the work or service is transferred, in return for either the right to operate the work or service that is the subject of the contract, or such a right in addition to a price for operating the work or service."[16]

Public procurement contracts typically cover the purchase of works, products, and services by public administrations and agencies, such as local authorities (which accounted for 50% of public procurement contracts in 2023[17]), the State, public institutions (such as Regional Health Agencies), and *sui generis* public entities, such as the French Agency for the Development and Promotion of Organic Agriculture (referred to as *Agence Bio* in French).

Before 2016, public administrations were allowed to include animal welfare as a requirement either for vendors' eligibility during the bidding process (during the application phase) or as the subject of a performance clause (in the contract itself as a mandatory clause). The 2018 reform, which resulted in the adoption of Article R2152-7 of the Public Procurement Code mentioned above, now explicitly allows public administrations to require compliance with animal welfare standards as a condition for participation in procurement procedures.

The range of public procurement contracts in which animal welfare requirements (as per Article R2152-7 of the Public Procurement Code) may apply is broad. Examples taken from a large variety of public procurement contracts concluded by local authorities with vendors include the supply of foodstuffs, introductory courses in dog training and

[14] Article L2 du titre préliminaire of the *Code de la commande publique* (Fr.). We translated.

[15] *Code de la commande publique* (Fr.), Article L1111-1. We translated.

[16] *Code de la commande publique* (Fr.), Article L1121-1. We translated.

[17] Caisse des Dépôts. 2024. Baromètre: la commande publique tirée par les collectivités locales. Available at: https://www.caissedesdepots.fr/actualites/barometre-la-commande-publique-tiree-par-collectivites-locales.

management in urban areas, the organization of sports and leisure activities for teenagers through youth services, the manufacture and installation of stables for pigs and sheep in the new premises at a university, the creation and production of equestrian shows, the installation of sheep fences, the provision of security guards with intervention dogs for night-time surveillance of the airport, and veterinary care and services for inter-municipal animal pounds.

Accordingly, food procurement is not the only context in which animal welfare standards are relevant. A wide range of public procurement contracts relate to animal welfare beyond the treatment of farmed animals used for food purposes, such as:

- Contracts that directly use animal body parts or products (supply of food, soap, work clothes, etc.);
- Contracts that indirectly use animals (supply of cleaning products, works, services and provisions using animal body parts or products tested on animals);
- Contracts aimed at limiting or eliminating animal populations (rat extermination, pigeon castration, etc.);
- Contracts related to live animals (grazing, guarding, leisure activities such as pony rides, work causing nuisance to animals, animal pounds, etc.).

2.2.2 Animal Welfare Provisions and Standards in Public Procurement Contracts

Examples of French municipalities incorporating animal welfare provisions into public procurement contracts include:

- Douai (2021): Incorporated an "animal protection" selection criterion weighing 5% of the final bidding score in its contract for the supply of cleaning products. This measure was taken by the councilperson responsible for the environment and animal welfare in the municipality;[18]
- Toulouse (2022): Included an animal welfare standard weighing 20% in a food supply contract[19];

[18] *Bulletin officiel des annonces des marchés publics* (Fr.), Avis No 22-89227.
[19] Ibid., Avis No. 22-118115.

- Strasbourg (2022): Included ambitious animal welfare clauses in a contract, requiring animal shelter services to walk dogs daily and have an agent trained in behaviorism on site, and imposing a €1000 penalty for any unjustified euthanasia[20];
- Auvergne-Rhône-Alpes Regional Council (2022): Incorporated an animal welfare standard weighing 25% for a food supply contract, for poultrymeat[21];
- Grenoble (2021): Adopted a clause requiring animal welfare requirements to be inspected on site, and a termination clause if an animal welfare requirement is not met;[22]
- Grand Besançon Métropole (2021): Asked its local officials to use selection criteria or performance clauses that incorporate animal welfare wherever possible.[23]

Despite the number and variety of these examples, there remain many obstacles to the inclusion of such provisions.

3 Barriers to the Inclusion of Animal Welfare Standards in Public Procurement Rules

Public administrations in France have traditionally underused their ability to include animal welfare standards in their public procurement rules under Article R2152-7 of the Public Procurement Code (Sect. 3.1.). This is mainly due to limited awareness of the provision and the complexity of applying it in practice, particularly in relation to animal welfare (Sect. 3.2.).

3.1 Article R2152-7 of the Public Procurement Code: A Provision Often Little Known

The establishment of animal welfare offices at local levels has contributed to greater awareness of the possibility of including animal welfare in public procurement rules as per Article R2152-7 of the Public Procurement

[20] Ibid., Avis No. 22-29226.

[21] Ibid., Avis No. 22-119234.

[22] Ville de Grenoble (2021), Séance du 14 juin 2021—Note de synthèse n°12-(25982). Environnement: Délibération Cadre—Condition et bien-être animal, orientations et plan d'action 2021–2026.

[23] Grand Besançon Métropole (February, 2021), Schéma de promotion des achats socialement et écologiquement responsables (SPASER) 2020–2026.

Code. For instance, several municipalities and local governments—such as Lyon, Grenoble, Strasbourg, Bordeaux Pessac, Douai, and Villers-sur-Mer—have developed more comprehensive approaches to animal welfare issues.

At the local level, the actors involved in integrating animal welfare into public contracts fall into two main categories:

- Elected officials, who drive the political impetus. Some may also sit as members of the Tender Committee, a body responsible for selecting the successful bidders for a public contract through a formal procedure, as required under EU law when the estimated value of the contract, excluding tax, exceeds the applicable threshold.[24]
- Local civil servants, who work within the different procurement departments or operational departments, which specialize in the execution of contracts.

The inclusion of new public procurement requirements also requires in-depth knowledge of the various departments and agencies involved and how they each interact with one another. More specifically, it is essential to understand the role of purchasing departments in the decision-making process and the discretionary powers and personal preferences of civil servants tasked with conducting the legal procedures necessary to incorporate new requirements.

3.2 *Varying Degrees of Societal Expectations Depending on the Administration and Sector*

Due to their direct and daily contact with citizens, local authorities are often subject to relatively strong societal pressure. However, certain public administrations—such as hospitals or prisons—face less scrutiny, as their material costs are relatively low and their activities attract less political attention from citizens, who are less exposed to their actions.

Societal expectations tend to focus on food purchasing policies, where impact on animals is more visible to the public. Pressure is particularly strong with respect to the treatment of egg-laying hens and cows—most often in the form of demands for sourcing eggs from free-range systems

[24] *Code général des collectivités territoriales* (Fr.), Article L1414-2.

and for beef from producers applying best animal welfare practices during transport or slaughter.

The attention paid by citizens to laying hens and cows also reveals that the inclusion of animal welfare standards in public procurement often depends on how the public perceives the treatment of animals. Yet these perceptions do not always reflect the true degree and scale of animal suffering. Additional farmed animals beyond cows and hens—which represent only a relatively small proportion of the total number of animals used for food—are also subjected to abuse in the food industry. Observations about societal expectations in the specific context of public procurement are consistent with economic research, which has shown that consumers are more willing to pay for the welfare of cows, hens, and chickens than for that of pigs and fish.[25] As a result, services that are not directly related to food, such as the supply of cleaning products or the performance of work, are not typically areas that the public associates with animal welfare.

3.3 The Complexity of Animal Welfare as a Concept

Animal welfare is a concept that can be difficult for stakeholders to fully grasp, which often hinders its systematic integration into procurement processes. For example, the empirical analysis conducted for this research showed that neither the "Five Freedoms" principles[26] nor the definition of animal welfare provided by the French food agency (ANSES)[27] is widely known. While the ANSES definition emphasizes the animal's own perspective of its welfare, stakeholders involved in public purchasing policies often overlook this point of view. Moreover, the concept of animal welfare is often conflated with animal care or animal protection.

The challenges stakeholders have in precisely identifying what the concept of animal welfare encompasses more generally stem from epistemological issues and the absence of a clear scientific and regulatory definition. Additionally, applications of the concept also vary depending on the animal species and the aspect of animal life under consideration, meaning that

[25] Espinosa, R. 2021. *Comment sauver les animaux*. Paris: Presses Universitaires de France, p. 67 (in French).

[26] Farm Animal Welfare Council (FAWC). 1979. Press statement.

[27] Anses. 2022. Bien-être des animaux_ quels critères scientifiques pour construire un étiquetage pertinent? Available at: https://www.anses.fr/fr/content/bea-etiquetage.

animal welfare cannot be reduced to a single measure, such as outdoor access for farmed animals.

As an academic discipline, animal welfare also suffers from compartmentalization from other disciplines, which tends to segment knowledge and limit opportunities for integrating animal welfare concepts into the legal disciplines, social sciences, and the humanities. The scientists at ANSES also formulated this observation in a 2018 opinion,[28] stating that "the topic of animal welfare is at the crossroads of many, sometimes contradictory philosophical and moral, scientific, technological and economic, regulatory and societal influences," despite the recent development of transdisciplinary educational programs. The issue of animal welfare in public procurement typically exemplifies the need to engage in transdisciplinary reflection in animal welfare science and law.

3.4 *The Technicity of Public Procurement Law*

The regulations governing public procurement are highly complex, requiring specific legal training. Because the rules largely rely on case law, public procurement law requires continuous monitoring, making it challenging to follow even for public procurement and animal welfare experts. As a result, relatively few people are aware that animal welfare standards can be used as a requirement when awarding contracts.

3.5 *The Proliferation of Standards in Public Procurement Rules*

Public procurement policies have also grown more complex as a result of new sustainable development standards that administrations are required to incorporate. One example is the standards introduced by the 2018 and 2022 EGALIM laws[29] in France, which require public administrations to include sustainable food standards in their procurement rules. Because animal welfare standards are not binding for administrations, public administrations thus tend to favor sustainability standards over animal

[28] Anses. 2018. Le bien-être animal en 8 questions. Available at: https://www.anses.fr/fr/content/anses-definition-du-bien-etre-animal. Accessed 30 April 2025.

[29] Loi n° 2018-938 du 30 octobre 2018 pour l'équilibre des relations commerciales dans le secteur agricole et alimentaire et une alimentation saine, durable et accessible à tous, *Journal officiel de la République française*, 1 November 2018.

welfare ones. And while the sustainable development requirements under the EGALIM law might appear consistent with animal welfare objectives, this is not always the case.

Since 2022, the EGALIM law has required public catering services to provide at least 50% sustainable, high-quality products, including at least 20% of organic products.[30] Non-organic products are considered "sustainable" if they carry one of the following:

- an official quality label, such as protected designation of origin, *Label Rouge,* protected geographical indication, organic farming, and eco-label;
- a promotional label (guaranteed traditional specialty, High Environmental Value, farm products, etc.);
- the Sustainable Fishing eco-label for seafood products.

However, these labels do not always include strong animal welfare standards.[31]

The 2018 EGALIM law also introduced two pilot programs, launched in April and November 2019, respectively. The first allowed voluntarily participating administrations to disclose the products used in the composition of menus in collective catering. The second program required all school cafeterias to offer a vegetarian menu at least once per week. The 2021 Climate and Resilience Law codified the weekly vegetarian menu requirement to make it permanent and further mandated that, beginning in 2024, 60% of meat and fish products must be sourced from sustainable and quality labels (100% for all public cafeterias). However, French legislation did not establish inspection mechanisms or penalties, meaning compliance has relied largely on societal pressure (such as feedback from parents of schoolchildren in the context of school cafeterias).

Beyond requirements specifically concerning food products, public administrations must also increasingly incorporate sustainable

[30] Ibid.

[31] On animal welfare standards in EU quality labeling scheme, see Di Concetto, A. and M. Friant-Perrot. 2021. Le bien-être animal et l'information des consommateurs. *Revue de l'Union Européenne* 651; Di Concetto, A. 2023. Farm Animal Welfare and Food Information for EU Consumers: Harmonizing the Regulatory Framework for More Policy Coherence. *European Journal of Risk Regulation* 1/2023; Di Concetto, A., and A.-S. Epstein. 2023. EU Consumer Information as a Tool to Regulate the Treatment of Farm Animals: Potential and Limits. *European Journal of Consumer Law*: 145–174.

development standards, as per Article L.3-1 of the Public Procurement Code (as amended by the 2021 Climate and Resilience Law): "Public procurement contributes to the achievement of sustainable development objectives, in their economic, social, and environmental dimensions, under the conditions defined by this code." This provision elevates sustainable development to a legal principle, alongside freedom of access to public procurement, equal treatment of bidders, and transparency in purchasing procedures.

4 Operational Challenges During and After Contracts Are Awarded

There are currently no guidelines for including animal welfare in public procurement rules, and little feedback exists on the experience of administrations and staff who have implemented these policies. Nevertheless, stakeholders—particularly elected officials, administrative staff, and economic operators—face a number of operational difficulties in the implementation of these animal welfare requirements.

4.1 Assessing Good Animal Welfare Practices

The objective of integrating animal welfare into public procurement requires the ability to assess the good animal welfare practices that economic operators claim to follow. This assessment generally relies on indicators developed by veterinarians and animal welfare scientists. These indicators focus both on the animal's environment and, increasingly, on the animal's own experience, so that "this assessment [reflects] as accurately as possible what animals actually experience."[32]

The use of indicators is relevant because French legislation, under Article R.2152-7 of the Public Procurement Code, requires bidding standards to be tied to the material scope of the contract, as well as to be non-discriminatory, sufficiently precise, and balanced. Animal welfare indicators must therefore be directly related to the services requested in the call for tender, applicable to all candidates without discrimination, and measurable and verifiable.

In the context of farm animal welfare, three types of indicators generally allow for the assessment of these requirements:

[32] Mounier, L (ed.). 2021. *Le bien-être des animaux d'élevage*. Versailles: Editions Quae.

- Environmental indicators, which assess the animal's environment;
- Animal-based indicators, which are based on the animals themselves, including behavioral, physiological, production, and health indicators. Since it is not currently possible to access the animal's feelings and mental state, welfare is measured using these indicators.
- Management indicators, which are based on production standards and evaluate the practices implemented, as well as the management processes and tools used in production.

Based on this typology, various stakeholders—scientists, nonprofits, food manufacturers—have designed standards that serve as a basis for assessing animal welfare. These standards may differ depending on the species, the production methods, and the age of the animals. For example, the Welfare Quality protocol[33] is a widely accepted protocol to measure animal welfare in the EU and includes environmental and animal-based indicators. However, its implementation in a commercial context can be challenging, as it generally requires a full day of field observation.[34]

Another example of a useful animal welfare indicator is the Boviwell protocol,[35] which was developed in France by the beef producers Interbev and *Confédération Nationale de l'Élevage* and *Centre National Interprofessionnel de l'Économie Laitière* as a tool to assess the welfare of cattle. Boviwell is based on the Welfare Quality protocol, the ANSES definition of animal welfare, and the Five Freedoms framework. These indicators have further been integrated into voluntary animal welfare guidelines for dairy producers, which were enacted as a response to the Bovine Spongiform Encephalopathy crisis.[36]

[33] Welfare Quality Network. Welfare Quality assessment protocols. Available at: https://www.welfarequalitynetwork.net/en-us/reports/assessment-protocols.

[34] Mounier, L. (ed.). 2021. Le bien-être des animaux d'élevage. Versailles: Editions Quae, p. 45.

[35] Interbev. 2024. Évaluer le bien-être animal dans les fermes bovines françaises. Available at: https://www.interbev.fr/wp-content/uploads/2024/01/131713-interbev_1500ex-4p-boviwell-hdef-web.pdf. Accessed 30 April 2025.

[36] France Terre Lait. 2022. La charte des bonnes pratiques d'élevage. Available at: https://www.terroirs-de-lait.com/wp-content/uploads/2022/03/Plaquette-CBPE_Cniel_02-2022.pdf.

Animal protection organizations have also enacted animal welfare indicators. For instance, Compassion in World Farming (CIWF) designed a methodology for assessing the "key indicators" of broiler welfare.[37]

However, from a public procurement perspective, a major obstacle with these protocols remains the verifiability of the good practices that producers claim to follow. Labels have partly remedied this limitation, as consumer protection law requires certifiers to guarantee the existence of good practices through implementation measures set out in product specifications.

4.2 *The Use of Labels*

When sourcing eggs, public administrations can use the method-of-production labeling system as a way to source eggs from humane egg production methods. This label is the only mandatory label that informs retail consumers about the treatment of farmed animals in the production chain, ensuring traceability of one aspect of animal welfare in the production chain. However, this label provides only partial information, since it applies solely to shell eggs and covers the treatment of egg-laying hens on farms, excluding other species and stages of egg production. Despite these limitations, this label is reliable, as producers must label all eggs according to specifications established by law.[38]

Public administrations may also rely on voluntary labels, including quality scheme labels and regulated claims referring to animal welfare. These labels and regulated claims can further facilitate the assessment of animal welfare policies in public procurement practices. In the Netherlands, for example, the Ministry of Infrastructure and the Environment has included some of the standards of the private, voluntary animal welfare label *Beter Leven*, created in 2007 by the animal welfare association *Dierenbescherming*, in its public procurement rules. Under this rule, the bidding process for public contracts is eligible only to vendors who can certify compliance with the label's entry level ("one star"), which has

[37] CIWF Agroalimentaire. POULETS DE CHAIR: Indicateurs clés de Bien-être Animal. Available at: https://www.agrociwf.fr/media/6760128/poulets-de-chair-indicateurs-bea-ciwf.pdf.

[38] Commission Delegated Regulation 2023/2465 of 17 August 2023 supplementing Regulation 1308/2013 of the European Parliament and of the Council as regards marketing standards for eggs, and repealing Commission Regulation 589/2008, 2023 O.J. (L 2023/2465).

higher requirements than the minimum legal standards. This public procurement practice has helped legitimize the label and facilitate its integration into the Dutch market.[39]

In France, the constitutional principle of free administration of local authorities[40] limits the government's authority to mandate public procurement practices. However, the French Constitution also provides that the law determines the fundamental principles of the free administration of local authorities.[41] In this context, the French Constitutional Council has confirmed the legality of certain obligations for local authorities, such as the introduction of a weekly vegetarian menu in school cafeterias in 2018.[42] Public procurement requirements set in national law are therefore legal in French law too, provided these requirements do not disproportionately undermine the principle of free administration of local authorities. The legislature may therefore require local authorities to include animal welfare labels in public procurement rules.

However, animal welfare labels also have shortcomings. There are no comprehensive, reliable animal welfare labels across industries or for specific sectors. In the food industry, only one dedicated label—the "Animal Welfare Label" created by the "Association Étiquette Bien-être Animal"—provides information on the welfare of broiler chickens, based on a set of 230 specifications.[43]

With such shortcomings, some existing labels, including EU quality schemes labels, do include animal welfare standards. This is the case for the EU organic label, as well as *Label Rouge* for certain animal products. Yet animal welfare specifications vary significantly between labels and within a single label across species, which may confuse buyers. For example, *Label Rouge* pork standards do not require outdoor access for pigs, leading to confinement that may in turn result in mutilations such as tail docking. Only the regulated claim "free-range" guarantees that pigs have outdoor access.

Compared to pork, the poultry welfare standards in the EU organic label and *Label Rouge* are more ambitious, although significant disparities

[39] Bismuth, R., et al. 2018. La concurrence des normativités au cœur de la labellisation du bien-être animal. *Revue Internationale de Droit Économique*: 369–393.

[40] *Constitution française du 4 octobre 1958* (Fr.), Article 72.

[41] Ibid., Article 34(13).

[42] Conseil constitutionnel, Décision n° 2018–771 DC du 25 octobre 2018.

[43] Étiquette Bien-Être Animal. Animal welfare label. Available at: https://www.etiquette-bienetreanimal.fr/. Accessed 25 April 2025.

remain. For example, organic broiler chickens must have outdoor access and a minimum area of 4 m^2 per chicken, as must *Label Rouge* "free-range" chickens. In contrast, *Label Rouge* "open air" requires only 2 m^2 per chicken.

When comparing these labels, the EU organic label appears most reliable because its specifications and control mechanisms are robust and codified in law. The label also has a strong presence in the food sector, making it easy to source large quantities of animal products. Additionally, the EU organic label includes animal welfare standards and specifications, as well as others related to sustainable development. In fact, EU regulations governing organic agriculture have gradually integrated the concept of animal welfare since the enactment of the EU organic legislation in 1991.[44]

With the 2007 reform of the EU organic laws, the pursuit of "high animal welfare standards" further became one of the main objectives of the EU organic labels.[45] A number of practices that cause pain in animals are therefore prohibited in organic food production, including the tethering of animals, the use of cages for poultry, force-feeding, and high stocking densities. Conversely, slow-growing breeds for poultry and outdoor access for mammals are mandatory.[46] The 2018 organic regulations further improved these standards by including new animals in their scope (rabbits, farmed fish, deer, and breeding sows) and by specifying certain standards, such as density in duck and turkey farms.[47]

[44] Council Regulation (EEC) No. 2092/91 of 24 June 1991, on organic production of agricultural products and indications referring thereto on agricultural products and foodstuffs, 1991 O.J. (L 198), p. 1.

[45] Council Regulation (EC) No. 834/2007 of 28 June 2007, on organic production and labeling of organic products and repealing Regulation (EEC) No. 2092/9, art 3(a)(iv).

[46] Duval, E., von Keyserlingk, M., and B. Lecorps. 2020. Organic Dairy Cattle: Do European Union Regulations Promote Animal Welfare? *Animals* 10: 1786; Di Concetto, A. 2023. Farm Animal Welfare and Food Information for EU Consumers: Harmonizing the Regulatory Framework for More Policy Coherence. *European Journal of Risk Regulation* 1/2023; Di Concetto, A., and A.-S. Epstein. 2023. EU Consumer Information as a Tool to Regulate the Treatment of Farm Animals: Potential and Limits. *European Journal of Consumer Law*: 145–174.

[47] Regulation (EU) 2018/848 of the European Parliament and of the Council of 30 May 2018 on organic production and labeling of organic products and Commission Implementing Regulation (EU) 2020/464 of 26 March 2020 laying down certain rules for the application of Regulation (EU) 2018/848 of the European Parliament and of the Council as regards the documents necessary for the retroactive recognition of conversion periods, the production of organic products and the information provided by Member States.

But while the EU organic label's specifications are stricter than current industry practices, these standards do not guarantee high levels of animal welfare, especially for animals other than poultry.[48] Furthermore, some EU organic rules remain unspecified, especially regarding the treatment of animals during transport and slaughter. EU case law has addressed these gaps only partially, such as with the 2019 ruling, which determined that meat certified organic had to come from animals that had been stunned at the time of killing.[49]

In the absence of comprehensive and reliable animal welfare labels, organic production remains, however, the most straightforward way to include animal welfare standards in public procurement policies. Furthermore, the organic label is especially relevant given the EGALIM law, which requires public administrations to source a minimum percentage of food products from organic production.

Beyond existing labels and certifications, the implementation of effective animal welfare requirements in public procurement policies requires a careful analysis of label specifications, so that the buyer can measure the impact that these specifications have on animals. In this regard, three initiatives can guide public administrations in their assessment work. In 2020, the nonprofit Info-Track published Guidelines for Commercial Communications on Animal Welfare,[50] aiming at better informing consumers. The animal protection nonprofit Welfarm also published a handbook[51] for public buyers analyzing selected label specifications. Similarly, the animal protection nonprofit Compassion in World Farming provides comparative tables of animal welfare and quality labels on its website.[52]

[48] Di Concetto, A. 2023. Farm Animal Welfare and Food Information for EU Consumers: Harmonizing the Regulatory Framework for More Policy Coherence. *European Journal of Risk Regulation* 1/2023.

[49] Case C-497/17, *Oeuvre d'assistance aux bêtes d'abattoirs (OABA) v Ministre de l'Agriculture et de l'Alimentation and Others*, Judgment of 26 February 2019.

[50] Association InfoTrack. 2021. Principes Directeurs relatifs aux communications commerciales en matière de bien-être animal. Available at: https://info-track.org/contenu/. Accessed 30 April 2025.

[51] WELFARM. 2021. Animal welfare vade mecum: proposal for animal welfare specifications in public procurement. Available at: https://appro-etica.fr/wp-content/uploads/2021/12/proposition_de_specifications_relatives_au_bien-etre_animal-1.pdf. Accessed 30 April 2025.

[52] Compassion in World Farming (CIWF). 2020. Comparison of existing certifications for farmed fish against CIWF criteria. Available at: https://www.agrociwf.fr/media/7441465/

One solution for public administrations is therefore to require a label (or equivalent to comply with non-discrimination principles) or to use the animal welfare label's specifications and integrate them into their own requirements, technical documentation, consultation rules, and award criteria.

4.3 *Monitoring Animal Welfare Requirements in Public Procurement Contracts*

Economic operators using labels rely on third parties, namely certification bodies (or similar), to verify the accuracy of claims. For economic operators certifying compliance with standards equivalent to those established in a label, public authorities must implement a verification system before and after the contract is awarded.

4.4 *The Proliferation of Purchasing Criteria*

Public procurement must meet multiple objectives apart from simply providing resources to public administrations: public procurement must be effective (by selecting the "most economically advantageous offer"), as well as "sustainable," "innovative," or "social," as indicated in Article L.2112-2 of the French Public Procurement Code. In addition, Article L.3-1 of the same code now requires public administrations to include sustainable development standards in their purchasing activities.

While animal welfare standards contribute to sustainability objectives, their inclusion can add complexity to public procurement practices already burdened by social and environmental criteria. Nevertheless, there are many ways to integrate animal welfare standards effectively in the short and medium term.

5 Potential Areas of Development and Improvement

Several avenues exist to mainstream animal welfare standards in public procurement policies, such as training decision-makers, promoting systematic analysis of animal welfare standards in contracts, improving the

ciwf-comparatif-certifications-par-rapport-aux-criteres-ciwf-juillet-20204.pdf. Accessed 30 April 2025.

traceability of the production and processing of animal-source food, developing humane sourcing methods, and controlling costs through overall costs or life-cycle analyses.

5.1 Training Key Players

It is essential that those drafting calls for tender, those who support them (consultants, project management assistants), and decision-makers (particularly elected officials) have thorough knowledge of both the rules and best practices for animal welfare in the specific context of public procurement policies. Yet, the majority of the key players, such as elected officials responsible for animal welfare or public procurement, too often lack the necessary knowledge about the integration of animal welfare objectives into public procurement policies. Training requirements for stakeholders responsible for purchasing policies could remedy this situation, while increased networking between officials and public purchasers could facilitate peer-to-peer exchanges, pool knowledge, and identify best practices.

5.2 Mainstreaming Animal Welfare Criteria in Public Procurement Purchasing Decisions

Public administrations should integrate animal welfare objectives into decision-making processes prior to drafting contracts, so as to incorporate animal welfare requirements even in contracts where the link with animals is not immediately apparent. Some municipalities— such as the cities of Grenoble, Bordeaux, Pessac, Villers-sur-Mer, Douai, Lyon, and Strasbourg—have already taken steps in this direction.

5.3 Providing Traceability on Animal Welfare

Labels make it easier for public purchasers to evaluate and monitor their public procurement policies compared to relying on animal welfare standards in isolation. There currently exist 24 voluntary labeling systems related to animal welfare conditions in the EU, such as the Animal Welfare Label in France and three labels issued by public bodies in Denmark (*Bedre Dyrevelfaerd*) and Germany (*Mehr Tierwohl*).[53] The proliferation of these

[53] See Di Concetto, A. 2021. Food Labeling and Animal Welfare: Research Note #2. The European Institute for Animal Law & Policy.

labels and their associated standards prompted the European Commission to propose harmonized EU food labeling in May 2020,[54] and again in 2025,[55] to increase transparency around how animals are raised, transported, and slaughtered. Until such a label is enacted, existing labels and other information should be made more transparent so that public purchasers can make more informed choices.

5.4 *Developing Sourcing Methods*

French law requires public purchasers to follow sourcing methods, meaning that purchasers must meet with economic operators prior to the establishment of contracts to exchange views, gain a better understanding of available solutions, inform economic operators of their requirements, and ultimately gain a precise understanding of the scope of future contracts.[56] In 2019, the French administration published guidelines to guide these exchanges.[57]

The 2016 reform further provided a regulatory framework for sourcing. However, because this practice was previously allowed and was the subject of the 2019 guidelines, public purchasers have implemented sourcing methods that differ from one administration to the next. Some administrations have refused to investigate sourcing methods, for fear of violating competition rules, though transparency in sourcing methods is encouraged by the French Ministry of Agriculture for catering and food services contracts. Sourcing methods thus represent an opportunity to familiarize buyers with the practice of requesting animal welfare commitments from prospective vendors.

5.5 *Analyzing, Controlling, and Justifying Costs*

The main barrier to integrating more humane standards in public procurement policies is the cost of humane products. While more humane

[54] European Commission. 2020. Farm to Fork Strategy for a fair, healthy and environmentally friendly food system. COM(2020) 381 final.

[55] European Commission. 2025. Vision for agriculture and food. COM(2025) 75 final.

[56] *Code de la commande publique* (Fr.), Article R.2111-1.

[57] Legal Affairs Directorate—Ministry of Economy, Finance and Recovery. 2019. Guide to good sourcing practices. Available at: https://www.economie.gouv.fr/files/files/directions_services/fininfra/Guide_sourcing.pdf.

products are often pricier, the true cost of these products also depends largely on the cost assessment methodology.

One key aspect of such methodology involves the difference between price and total cost. On one hand, price is the face value, in currency, of a good or service at a given moment. It is expressed in monetary units, i.e., in euros for French public entities. On the other hand, the total cost is the economic value of the efficiency of a purchase. In sustainable development, the total cost approach aims to anticipate constraints (particularly in terms of production, operation, use, and maintenance). The price is therefore only a part, often a minimal part, of the total cost. Total cost analysis is a decision-making tool that is used in particular for work contracts.[58]

Another assessment tool that is relevant, although less frequently used, is life-cycle analysis for products and services. These life-cycle analyses, which take environmental externalities into account, consider both the negative and positive externalities of a product or service regarding the product's impact on animals and their treatment.

Quantifying these costs is necessary for public administrations to be able to consider life-cycle costs as an award requirement.[59] Yet, determining and verifying the monetary values attributed to the costs associated with good animal welfare practices can be challenging. To overcome this difficulty, public authorities and professional sectors could develop specifications and scoring systems to guide public purchasers. Such systems already exist, for example those measuring energy consumption levels or greenhouse gas emissions for certain products. In the case of animal welfare, calculation methods could be based on the following criteria: percentage of lifetime spent indoors, space per animal, or method of slaughter.

6 Conclusion

The analysis of public procurement in France shows the central role of economic law in advancing animal welfare. The transformative potential of this area is all the greater given that public procurement accounts for a considerable share of demand in certain markets. Through public procurement, public authorities have the opportunity to bring about production and consumption practices that are more consistent with animal welfare

[58] Ministère de l'Ecologie, de l'Energie, du Développement durable et de l'Aménagement. Le coût global. Available at: http://www.coutglobal.developpement-durable.gouv.fr/.

[59] *Code de la commande publique* (Fr.), Article R.2152-9.

objectives. Despite the existence of robust legal bases in EU and national law for integrating animal welfare into contract selection or contract performance clauses, these possibilities remain under-exploited. The obstacles identified relate both to the technical nature of public procurement law and to the complexity of quantifying and defining animal welfare itself. Nevertheless, local initiatives promoting more ethical public procurement demonstrate that concrete progress is possible when political will aligns with ethical procurement objectives.

The Protection of Endangered Species in EU and International Trade Law

Sophie Grosbon

1 Introduction

The Washington Convention on International Trade in Endangered Species of Wild Fauna and Flora[1] (CITES) was adopted on 3 March 1973, and came into force in 1975. Although resolutions adopted by the Conference of the Parties (CoP) are non-binding, these resolutions supplement the Convention by providing harmonized interpretations and guidance for its implementation. CITES exemplifies a liberal–structural approach to environmental issues.[2] Rather than prohibiting trade outright to address the risk, of extinction of certain wild species, the Convention regulates such trade, to prevent illicit and unregulated trafficking (liberal

[1] *United Nations Treaty Series*. 1976. Vol. 993, I-14537, p. 272 and following.

[2] On the liberal, structural, and systemic discourses on the relationship between environment and development: Morin, J.-F. and A. Orsini. 2015. Chapter 5. Sommets des Nations Unies: entre environnement et développement. In *Politique internationale de l'environnement*, 133–156. Paris: Presses de Sciences Po (in French).

S. Grosbon (✉)
Department of Law, University of Paris Nanterre Law School, Nanterre, France
e-mail: sgrosbon@parisnanterre.fr

A.-S. Epstein, A. Di Concetto (eds.), *Animals in EU Economic Law*, The Palgrave Macmillan Animal Ethics Series,
https://doi.org/10.1007/978-3-032-16171-0_9

approach) and to support the development needs of local populations (structural approach).

Since 8 July 2015, the EU has been a Party to CITES, which currently has 185 Parties.[3] However, the first EU regulation in the field of wildlife protection dates back to the early 1980s. Regulation 3626/82 on the implementation of CITES in the Community[4] aimed to implement the Convention's rules uniformly across the then-EU, the European Economic Community (EEC), at a time when not all EEC Member States were parties to CITES. Indeed, the principle of free movement of goods on the EEC market hindered the effective enforcement of CITES in those Member States that had ratified it.

The advent of the internal market, the growing need for better harmonization in the implementation of the Convention, the resolutions of the CITES' CoP, and the adaptation to new methods of trade control and species conservation gradually led to a restructuring of the EU's regulatory framework in the early 1990s.[5] In 1996, the EU adopted Regulation 338/97 on the Protection of Species of Wild Fauna and Flora by Regulating Trade Therein (EU Wildlife Trade Regulation),[6] which was later supplemented and further specified by additional regulatory instruments, including Regulation 865/2006 (Implementing Regulation)[7] and Regulation 792/2012 (Permit Regulation).[8] The EU regulations broadly reflect the spirit of CITES while explicitly detailing how its provisions are to be implemented. They frequently incorporate non-binding resolutions

[3] It was the entry into force in 2013 of the Gaborone Amendment, adopted in 1983, that authorized regional economic organizations—and hence the European Union—to become Parties to CITES. Since then, under Article XXI§5 of the Convention, the European Union has exercised the right to vote in place of the Member States at the COP, in its areas of competence, with a number of votes equal to the number of Member States.

[4] Council Regulation (EEC) No 3626/82 of 3 December 1982 on the implementation in the Community of the CITES, *OJEC* L384/1 of 31 December 1982.

[5] European Commission and TRAFFIC. 2020. *Reference Guide to European Union Wildlife Trade Regulations*. Belgium, Brussels, pp. 122–123.

[6] Council Regulation (EC) no. 338/97 of 9 December 1996, *OJEC* L61 of 3 March 1997, p. 1.

[7] Commission Regulation (EC) No. 865/2006 of 4 May 4 2006 laying down detailed rules for the application of Regulation No. 338/97, *OJEU* L166 of 19 June 2006, p. 1. This regulation has been substantially amended on several occasions.

[8] Regulation (EU) No. 792/2012 of 23 August 2012 laying down rules on the form of permits, certificates, and other documents provided for in Regulation 338/97 and amending Regulation 865/2006, *OJEU* L242 of 7 September 2021, p. 13.

from the CoP—sometimes using their wording verbatim—thus making them binding in EU law. In some areas, the regulations contain provisions that diverge from those of CITES to reflect the unique nature of European integration or to enhance the effectiveness of efforts to combat illegal trade. The new EU Action Plan against Wildlife Trafficking, adopted in November 2022, also notes that "the EU functions as a hub for global wildlife trafficking, and has a key role to play in the fight against it."[9]

The main objective of both CITES and the EU Wildlife Trade Regulations is to regulate the highly lucrative *cross-border trade* in certain wild species, with the aim of preventing overexploitation that could threaten their survival. Unregulated trade remains the default rule, with restrictions constituting the exception, as reflected in both the territorial and substantive scope of international and EU law.

The territorial scope of both CITES and EU Wildlife Trade Regulations primarily covers *cross-border trade*, although notable differences exist between the two regimes. CITES applies exclusively to international trade, establishing conditions for the import and export of specific species.[10] It does not extend to species—even endangered ones—when traded within domestic markets. By contrast, EU law is more comprehensive, as it covers both intra-EU trade and trade between the EU and third countries. Moreover, the EU appears to prohibit the commercialization of the most endangered species even at the national level, although numerous exceptions to this principle significantly undermine its effectiveness.[11]

The substantive scope of both CITES and the EU Wildlife Trade Regulations excludes *non-endangered species*. Consequently, the trade in such species is generally unrestricted. The Convention only governs trade in species that are threatened with extinction (listed in Appendix I), species likely to become threatened without regulation (Appendix II), and nationally protected species vulnerable to overexploitation through international trade (Appendix III).[12] Annexes A, B, and C to the EU Wildlife Trade Regulation broadly correspond with CITES Appendices I, II, and III, respectively. However, the EU Annexes also include species subject to

[9] Communication from the Commission to the European Parliament, the Council, the European Economic and Social Committee and the Committee of the Regions. 2022. *Revision of the EU Action Plan against Wildlife Trafficking*. COM(2022) 581 final, 9 November, p. 2.

[10] CITES, Article I c.

[11] EU Wildlife Trade Regulation, Article 8.

[12] CITES, Article II.

stricter protection measures adopted by the Union[13] and exclude species for which Member States have entered reservations under CITES.[14]

As a result, CITES only restricts cross-border trade in endangered species, which is subject to "strict" or even "particularly strict" regulatory requirements.[15] Trade in a specimen of a species listed in Appendix I requires both an export and an import permit,[16] whereas trade in a specimen of a species listed in Appendix II requires only an export permit.[17] The EU Wildlife Trade Regulation is more stringent in this respect, as it requires both an export and an import permit for species listed in Annexes A and B.[18] Both CITES and the EU regulations mandate that competent authorities issue permits only if traders meet certain conditions—such as proving that the specimen was legally acquired, and demonstrating compliance with applicable transport and care standards at the destination. Among these conditions, the issuance of a "non-detriment finding" is a key element: a national scientific authority must confirm that the trade in the specimen will "not be detrimental to the survival of th[e] species," as required by CITES—or more strictly, under the EU Wildlife Trade Regulation, that the trade "will not have a harmful effect on the conservation status of the species or on the extent of the territory occupied by the relevant population of the species."

This regulated trade, governed by a core requirement relating to the survival or conservation of the species, does not fall within the realm of animal ethics, but rather that of environmental ethics, which, fundamentally, "is not concerned [...] with the welfare of individual wild animals, but [...] with the protection of population dynamics compatible with the objective of maintaining biodiversity."[19] According to this logic, wild animals are not seen as beings beyond the reach of commerce or appropriation. On the contrary, they are considered subject to human

[13] Article XIV of CITES authorizes Parties to adopt stricter domestic measures to protect endangered species.

[14] EU Wildlife Trade Regulation, Article 3. The regulation also includes a fourth annex, Annex D, listing "species not listed in Annexes A to C whose volume of Community imports is such as to justify monitoring."

[15] CITES, Article II.

[16] CITES, Article III.

[17] CITES, Article IV.

[18] EU Wildlife Trade Regulation, Article 4.

[19] Epstein, A.-S. and A. Di Concetto. 2022. Quelle place pour le bien-être des animaux d'élevage en droit de l'environnement. In *Bien-être et normes environnementales*, ed. Isabelle Michallet, p.124. Paris: Mare et Martin (in French).

control—captured, owned, or otherwise regarded as property. CITES does not establish any obligation to respect an animal's desires, free development, or private and family life, let alone any right for animals to flourish. In less anthropocentric terms (setting aside references drawn from human rights law), CITES does not impose any obligation to respect the animal's wild state.[20] In the wild, an individualized animal, regarded in isolation from its species, is of no concern to CITES, which views it solely as a potential commodity or a reservoir of genetic and reproductive material (I). It is only once the animal is held in captivity, commercialized, and alienated that its welfare is even minimally addressed by CITES and EU regulations (II).

2 The Appropriation of Wild Animals

CITES and EU regulations cover "wild" animal species, yet neither defines what "wild" actually means. The term does not exclusively refer to animals living in their natural habitats, as both regimes also apply to animals in captivity. A wild animal seems to be rather one that is not typically subject to domestication. But the notion of whether an animal is "meant" to be domesticated is deeply shaped by cultural perceptions. Ultimately, CITES was driven by the concern that international trade, particularly the removal of animals from their natural environments, was pushing some species toward extinction.

Nevertheless, the Convention does not prohibit the appropriation of animals; it aims to regulate it. This commodification of animals is typically justified by appealing to the essential needs of local populations and the economic benefits they expect to derive from it (A). Yet such a supply of wild animals could not flourish without corresponding demand—a demand that is often far removed from any vital necessity and driven primarily by consumer gratification (B).

2.1 Wild Animals as Vital Economic Resources

The impetus for regulation established by CITES is twofold. On the one hand, the Parties to the Convention recognize the need to protect "wild fauna and flora" as an "irreplaceable part of the natural systems," valued

[20] On the contrary, respect could be established as a principle, subject to limited fundamental exceptions.

for their beauty and variety.[21] On the other, they also acknowledge "the ever-growing value of wild fauna and flora from aesthetic, scientific, cultural, recreational and economic points of view."[22] However, such an anthropocentric and utilitarian rationale,[23] by itself, is insufficient to justify a legal regime based on trade regulation; it could just as readily have supported a prohibition-based model. The Parties' decision to regulate rather than ban trade reflects a compromise between efforts to prevent the extinction of certain species and the economic interests of local and Indigenous communities—primarily, though not exclusively, in developing countries—who would rely on this trade for their livelihood.[24]

A 1992 CoP resolution entitled "Recognition of the Benefits of Trade in Wildlife"[25] reinforces this interpretation, noting that most wild species protected by CITES are found in developing countries, and "that commercial trade may be beneficial to the conservation of species and ecosystems" because "the returns from legal use may provide funds and incentives to support the management of wild fauna and flora to contain the illegal trade." In this resolution, the Parties also recognize that trade can benefit "the development of local people," as regulated trade forms part of the broader framework of "sustainable use," which, in turn, "provides an economically competitive land-use option." In this light, the CoP views legal and regulated trade as a legitimate tool for sustainable species conservation. The resolution further suggests that prohibiting such use could be counterproductive, as it may encourage illicit trafficking, poaching, and transnational organized crime.

This 1992 CoP Resolution was amended in 2004 to stress that implementation of CITES-listing decisions should take into account "potential impacts on the livelihoods of the poor" before listing a species in the CITES Appendices and potentially triggering trade restrictions.[26] The "CITES and Livelihoods" resolution, adopted in both 2013 and 2019, addresses this concern by acknowledging that while "the implementation

[21] CITES, preamble, first subparagraph.

[22] Ibid., second subparagraph.

[23] Peters, A. 2020. *Animals in International Law*. RCADI, vol. 410, p. 144. The Hague: Hague Academy of International Law.

[24] Maljean-Dubois, S. 2021. *Le Droit international de la Biodiversité*. RCADI, vol. 407, p. 172; Cooney, R. and M. Abensperg-Traun. 2013. Raising Local Community Voices: CITES, Livelihoods and Sustainable Use. *RECIEL* 22 (3): 301–310.

[25] CITES. 1992 (revised 2004). Resolution Conf. 8.3 (Rev. CoP13), "Recognition of the Benefits of Trade in Wildlife."

[26] Ibid.

of CITES listings may enhance livelihoods by delivering long-term species conservation and reducing unsustainable and illegal trade," such listings may also "[restrict] access to income, employment and other resources, such as food, materials and medicines" for "rural communities."[27] The EU 2022 Action Plan Against Wildlife Trafficking reflects the same perspective, stating that "well-managed trade in wildlife can actively benefit species conservation, creating incentives for local communities to protect wildlife resources in their environment, while making a sufficient and sustainable living."[28] From this standpoint, the regulation of wildlife trade primarily seeks to meet the essential needs of local communities—in contrast with a prohibition-based regime, which may be perceived as reproducing post-colonial dynamics in ecological governance.

There are, however, alternatives to the legalization of wildlife trade that can also safeguard the basic needs of vulnerable populations. The CoP has acknowledged the importance of developing income sources independent of wildlife trade in its resolutions from 1992, 2013, and 2019. For instance, the 1992 resolution (albeit incidentally) highlights the "enormous importance" of "the aesthetic, scientific, cultural, recreational and other largely non-consumptive uses of wild fauna and flora."[29] Similarly, the 2019 resolution on "CITES and Livelihoods" refers explicitly to the Sustainable Development Goals (SDGs), and in particular to SDG 15 and target 15.c, which calls for "enhance[d] global support for efforts to combat poaching and trafficking of protected species, including by increasing the capacity of local communities to pursue sustainable livelihood opportunities."[30] The "CITES and Livelihoods" resolution also promotes "mitigation strategies" aimed at reducing the impact of regulated trade on rural communities. These include providing "assistance, including short-term financial support, to rural communities most severely affected by the implementation of the CITES-listing decisions" and "promoting alternatives to rural communities to enhance the effective implementation of CITES-listing decisions, for instance [...] income-generation approaches,

[27] CITES. 2013. Resolution Conf. 16.6 (Rev. CoP18), "CITES and Livelihoods."

[28] Communication from the Commission to the European Parliament, the Council, the European Economic and Social Committee and the Committee of the Regions. 2022. *Revision of the EU Action Plan against Wildlife Trafficking*. COM(2022) 581 final, 9 November, p. 2.

[29] CITES. Resolution Conf. 8.3 (Rev. CoP13), "Recognition of the Benefits of Trade in Wildlife." Emphasis added.

[30] SDGs, https://sdgs.un.org/goals/goal15#targets_and_indicators.

such as payment for ecosystem services, sustainable tourism, employment in eco-tourism or as game wardens."[31]

The EU also recognizes the need to develop alternatives to income derived from wildlife trade, and therefore prioritizes efforts to "prevent[...] wildlife trafficking and address[...] its root causes" by "strengthen[ing] engagement of indigenous peoples and local communities in the management and conservation of wildlife, including through support for the development of sustainable livelihoods in source countries."[32] An example of an EU-supported measure in this regard is the NaturAfrica Initiative, which identifies "key landscapes for conservation and development," aiming "to create jobs, improve security and sustainable livelihoods, while preserving the ecosystems and wildlife that are vital to all."[33]

Proposed alternatives include eco-tourism and other sustainable tourism activities, even if they often reinforce a model in which local populations provide services to wealthy consumers. Still, such strategies underscore the possibility of exploring options beyond regulated trade to address the urgent need to combat poverty—especially given that this market exists only because consumers are willing to engage in it, often regardless of any genuine need.

2.2 *Wild Animals as Status Symbols*

CITES and EU Wildlife Trade Regulations do not present the trade in endangered species as essential to consumers. However, beneath the surface of these texts, two underlying imperatives emerge: wildlife as a reservoir of scientific and medicinal resources (2.2.1.) and the satisfaction of a demand for wild animals driven largely by ostentation or pure leisure (2.2.2.).

2.2.1 *Wild Animals as Source of Scientific and Medicinal Resources*

Based on the Convention and the EU regulations, the only fundamental interest that may justify the export of wild species is non-commercial

[31] CITES. 2013. Resolution Conf. 16.6 (Rev. CoP18), "CITES and Livelihoods."

[32] Communication from the Commission to the European Parliament, the Council, the European Economic and Social Committee and the Committee of the Regions. 2022. *Revision of the EU Action Plan against Wildlife Trafficking*. COM(2022) 581 final, 9 November.

[33] Ibid., p. 2.

exchange between scientific institutions. At both the international and the EU levels, this scientific and educational interest justifies an exemption from the strict regulations governing the trade in species listed in the various annexes. This exemption applies only to dead animals, specifically "preserved, dried or embedded museum specimens."[34] In addition, the EU Wildlife Trade Regulation prohibits the use of species listed in Annex A for commercial gain, but provides an exemption for specimens "required under exceptional circumstances for the advancement of science or for essential biomedical purposes."[35] This specific exemption applies to live animals, in accordance with Directive 86/609/EEC on the Protection of Animals Used for Experimental and Other Scientific Purposes.[36]

Apart from exemptions granted for scientific research, only one resolution from the CoP, Resolution 10.19, addresses a fundamental need on the demand side, specifically related to "traditional medicines."[37] This resolution endorses the World Health Organization's recognition of "the importance of traditional medicines to the world's medicinal security and that millions of people depend on these medicines for primary health care." In line with broader public health objectives, Resolution 10.19 highlights the dangers of overexploiting endangered species to meet this demand and urges countries to comply with CITES regulations to ensure sustainability. It also encourages the development of alternative sources — including synthetic substitutes, products derived from less-threatened species, or, with proper safeguards, those obtained through artificial reproduction or captive breeding.

If scientific research and traditional medicine were the sole drivers of trade in threatened species, the issue could likely be contained.[38] However, CITES, its resolutions, and related EU regulations say little about the underlying nature of the demand that fuels this market. In reality, much of this demand appears to be motivated less by necessity and more by displays of wealth or personal gratification, far removed from any essential needs.

[34] Article VII(6) of CITES and article 7(4) of the EU Wildlife Trade Regulation.

[35] Article 8(3) (e) of the EU Wildlife Trade Regulation.

[36] Ibid.

[37] CITES. Resolution Conf. 10.19 (Rev. CoP14), "Traditional Medicines."

[38] See below, 3.1 Captive-Bred Animals as a Substitute for Wild Animals.

2.2.2 *Wild Animals as Status Symbols and Personal Pleasure*

Article VII(3) of CITES further provides an exemption to the strict regulation on trade in wild animals for "personal or household effects." This term includes specimens acquired legally for personal, non-commercial use, and transported either in personal luggage or as part of a household relocation.[39] The CoP recommends applying this exemption to both dead and live animals.[40] As a result, under CITES—and consistent with the CoP resolution[41]—the cross-border movement of personally owned live animals is exempt from the permit requirement and is instead only subject to a certificate of ownership.[42]

However, the purpose of this exemption is not to give tourists unrestricted freedom to acquire exotic specimens, as such practices would undermine the core objectives of the Convention. The scope of the personal or household effects exemption varies depending on the applicable appendix. The exemption applies to Appendix I specimens that were acquired in their owner's country of residence. It does not apply if the specimen was purchased abroad and imported into the owner's country of permanent residence. The exemption also does not apply to Appendix II specimens that were acquired abroad, taken from the wild, and imported into the country of residence. However, if the specimen was not taken from the wild, but was instead acquired abroad, it may qualify for the personal effects exemption.

The EU Wildlife Trade Regulation is stricter than CITES in several respects. First, although the Regulation also provides an exemption for "personal or household effects," this exemption is limited to dead animals only.[43] Second, consistent with the CoP resolution, private individuals moving live animals across borders are exempt from permit requirements and must instead provide only a certificate of ownership. However, under EU law, such certificates are granted only for specimens that have been born and bred in captivity, or for those acquired or introduced into the

[39] CITES. Resolution Conf. 13.7 (Rev. CoP17), "Control of Trade in Personal and Household Effects."

[40] CITES. Conf. 10.20, "Frequent cross-border movements of personally owned live animals."

[41] Ibid.

[42] The certificate of ownership is designed to prevent commercial exports from being concealed as personal effects.

[43] EU Wildlife Trade Regulation, Article 7(3).

Union before their inclusion in the protected annexes.[44] For all other live animals kept for personal purposes, the full set of trade restrictions applies. No such limitation exists under CITES.

Third, unlike CITES, the EU regulation does not allow the personal or household effects exemption for the export of specimens listed in Annexes A and B from the Union.[45] Imports of personal and household effects are also subject to stricter rules under EU law than under CITES. When a specimen listed in Appendix A is introduced into the EU for the first time, EU residents or individuals taking up residence in the Union cannot benefit from the personal or household effects exemption. Furthermore, while the first introduction of a specimen listed in Annex B does not require an import permit, it does require an export permit—regardless of the place of capture.[46]

However, the personal or household effects exemptions implicitly authorize, both under international and EU law, the hunting of "trophies" in the country of origin for the purpose of export to the hunters' country of residence. EU Implementing Regulation 865/2006 explicitly includes "hunting trophies" among the personal or household effects eligible for exemption, provided specific conditions are met.[47] In principle, cross-border trade in species listed in Annex A of the EU Wildlife Trade Regulation, and Appendix I of CITES, remains subject to both export and import permit requirements. For species listed in Appendix II, however, Article VII(3) of CITES exempts even the basic export permit requirement when the specimen was acquired outside the wild. In contrast, trade in hunting trophies from species listed in Appendix B only requires an export permit. Nevertheless, EU law does not allow the use of the personal and household effects exemption for the first introduction into the Union of hunting trophies from certain Annex B species "for which there are concerns as to the sustainability of trade in hunting trophies or for which there are indications of significant illegal trade."[48] In such cases, both an import and an export permit are required. Additionally, the

[44] European Commission. 2006. Commission Regulation (EC) No 865/2006 of 4 May 2006 laying down detailed rules concerning the implementation of Council Regulation (EC) No 338/97 on the Protection of Species of Wild Fauna and Flora by regulating trade therein, Article 37.

[45] Ibid., Article 58(2).

[46] Ibid., Article 57(2) and 57(3).

[47] Ibid., Article 57(1).

[48] Recital 3 of Regulation (EU) 2015/870 amending Regulation (EC) No 865/2006.

European Commission has imposed an import ban on certain hunted species, including their trophies.[49]

Despite the narrowly defined exemptions, the international and EU frameworks governing hunting trophies starkly illustrate the futility of the demand that fuels the trade in endangered species. The trophy itself functions as a symbol of triumph over an animal and reflects Bourdieu's concept of social distinction.[50]

The efforts undertaken by CITES bodies to reduce demand for products derived from endangered species underscore a range of underlying consumption drivers: the craving for exotic meats, the prestige associated with owning rare animal collections, the conspicuous display of luxury wildlife goods, and the sentimental value of tangible travel souvenirs. These investigations reveal distinct consumer profiles—for instance, ultra-wealthy businessmen who assert their social status by acquiring cheetahs as status symbols, or urban middle- and upper-class women drawn to ivory items that evoke elegance, style, and a sense of refined taste.[51]

Ultimately, the only form of personal satisfaction explicitly condemned by CITES is the use of an imported specimen for "primarily commercial purpose" when the species is listed as threatened with extinction (Appendix I). Trade for "purely private use"[52] is therefore not prohibited, as long as the conditions of the dual permitting system, namely, valid export and import permits, are fulfilled. The EU Wildlife Trade Regulation reaffirms the ban on primarily commercial use, applying it to both import and export of Appendix A specimens within and beyond the Union.[53] However, it goes further than CITES by prohibiting, within the internal market, the purchase, sale, or any use for commercial gain involving such specimens.[54] This means that even private use following an initial purchase could be restricted. Nonetheless, there are several exceptions to this general

[49] EU Wildlife Trade Regulation, Article 4(6), p. 67.

[50] Bourdieu, P. 1979. *La distinction, Critique sociale du jugement*. Paris: Les Éditions de Minuit (in French).

[51] Secretariat of the CITES. 2021. Draft Guidance for CITES Parties to Develop and Implement Demand Reduction Strategies to Combat Illegal Trade in CITES-listed Species, pp. 10–13. Available at: https://cites.org/sites/default/files/eng/prog/DR/CITES_Guidance_Demand_Reduction.pdf.

[52] CITES. Resolution Conf. 5.10 (Rev. CoP19), "Definition of the Term 'Primarily Commercial Purposes.'"

[53] EU Wildlife Trade Regulation, Article 4(1)(d) and Article 5(2)(c).

[54] EU Wildlife Trade Regulation, Article 8(1).

prohibition, including specimens legally introduced into the EU "for purposes which are not detrimental to the survival of the species concerned,"[55] as well as those originating from a Member State and lawfully taken from the wild.[56] Such exceptions create a legal pathway for purely private appropriation. Finally, trade in species listed in Appendix II of CITES or Annex B of the EU Wildlife Regulation is not subject to the same restriction on commercial use. As a result, personal satisfaction for buyers of these species—whose conservation status could become endangered without strict regulation—can be entirely profit-driven or speculative.

An animal may therefore be taken from the wild, as long as it is lawfully acquired and accompanied by a non-detriment finding confirming that its removal does not threaten the survival of its species. It can then be subject to cross-border trade, motivated both by the expected economic benefits for local communities and the purely personal satisfaction of the end consumer. Little attention is paid to the animal's wild state. International and EU regulations show concern for the animal's welfare only once it has been brought into captivity—as if such concern were relevant only in the context of its domesticated relationship with humans, rather than in honoring its existence as a wild being, free from human interference.

3 The Minimal Protection of Captive Animals

CITES and the EU regulations view captive breeding as a viable alternative to taking specimens from the wild[57]—one that can satisfy both seller and buyer without endangering the survival of the species. Captive-bred animals are thus often regarded as second-class specimens, or as creatures whose entire existence is oriented toward serving human needs and desires.[58] When this utilitarian framing does not apply, such animals are

[55] EU Wildlife Trade Regulation, Article 8(3)(c).

[56] EU Wildlife Trade Regulation, Article 8(3)(h).

[57] CITES. Resolution Conf. 17.7 (Rev. CoP19), "Review of Trade in Animal Specimens Reported as Produced in Captivity."

[58] See, for example, Resolution 5.10 (Rev. CoP19), which interprets the "exceptional circumstances" allowing the trade in Appendix I species (article II(1)) as authorizing trade in specimens taken from the wild only if "the importer has first demonstrated that: a) he has been unable to obtain suitable captive-bred specimens of the same species." Additionally, see Article 8(3)(e) of the basic regulation, which derogates from the ban on commercial exploitation of Appendix A species in cases of scientific or biomedical necessity when "specimens of the species born and bred in captivity are not available."

instead presented as tools of species conservation: either by contributing to the reproduction of endangered species, or by forming an "insurance population"[59] held in captivity to guard against extinction in the wild (Sect. 3.1.). The living conditions in breeding facilities receive little attention when these facilities are the source of traded specimens. However, they can affect the legality of trade when they serve as the destination. In practice, animals that are captive-bred or taken from the wild for trade receive only minimal consideration, with regulatory oversight limited to their transport conditions and future housing (Sect. 3.2).

3.1 *Captive-Bred Animals as a Substitute for Wild Animals*

Captive breeding has been promoted as an alternative to wild harvesting, seen as posing a lower risk to species survival and therefore justifying certain exemptions from the strict regulations governing trade in endangered species (Sect. 3.1.1). However, documented cases of fraud have prompted increased scrutiny and tighter controls over these breeding operations (Sect. 3.1.2).

3.1.1 Exemptions From the Strict Regulation of Trade

The CoP of CITES defines captive-bred specimens as "born or otherwise produced in a controlled environment,"[60] or more specifically, an environment "manipulated for the purpose of producing animals of a particular species," equipped with physical barriers to prevent the entry and escape of specimens, and "which may include, but are not limited to: artificial housing; waste removal; health care; protection from predators; and artificially supplied food."[61] Specimens bred for the purpose of producing other animals are included, under both international and EU law, among the exemptions from the strict trade controls applicable to endangered species.

Under Article VII, paragraph 4 of CITES, captive breeding "for commercial purposes" allows specimens that would otherwise fall under Appendix I to be treated under the more flexible legal regime of Appendix

[59] Communication from the Commission. 2022. Guidance document on animals bred in captivity under the EU Wildlife Trade Regulation, OJEU C 306/2 of 11 August 2022, p. 3.

[60] Conf. 10.16 (Rev. CoP19), "Specimens of Animal Species Bred in Captivity"; Implementing Regulation, paragraph 2(b).

[61] Conf. 10.16 (Rev. CoP19), "Specimens of Animal Species Bred in Captivity"; Implementing Regulation, paragraph 1(d).

II. As a result, these specimens may be exported for commercial purposes without an import permit. Paragraph 5 of Article VII is more ambiguous, and the way it interacts with Paragraph 4 is not immediately clear. Article VII(5) appears to apply to captive-breeding operations involving Appendix I specimens when the purpose is *non-commercial*—since commercial breeding is explicitly addressed in paragraph 4. It likely also covers both commercial and non-commercial captive breeding of specimens listed in Appendices II and III. In these cases, a certificate issued by the Management Authority confirming that the specimen was bred in captivity may substitute all other permits and certificates. Once this captive-breeding certificate is granted, trade in these specimens—whether for commercial or non-commercial purposes—is permitted without the need for export or import permits, a non-detriment finding, or oversight regarding transport or housing conditions at the destination.

The exemption in EU law is more straightforward: captive-bred specimens of Annex A species are treated as Annex B specimens, regardless of whether the breeding is for commercial or non-commercial purposes.[62] The prohibition on the use of Annex A species for profit within the internal market is subject to an exemption for specimens that are born and bred in captivity.[63]

Captive breeding was originally intended to prevent the removal of animals from their natural habitats and to enable trade without endangering wild populations. However, the living conditions of animals within breeding facilities fall outside the scope of wildlife trade regulations. At best, a model application form for the registration of a captive-breeding operation—proposed by the CITES CoP as an annex to a resolution—requires applicants to indicate, among roughly fifteen listed items, how breeding is carried out "at all stages to ensure that animals are treated in a humane (non-cruel) manner."[64]

The captive breeding of inherently wild (i.e., non-domesticated) animals is consistent with the logic of the regulated trade system established by CITES and EU law. Initially conceived as a way to avoid wild harvesting deemed harmful to the survival of endangered species, captive breeding has, over time, given rise to widespread fraud and numerous

[62] Wildlife Trade Regulation, Article 7(1).

[63] Wildlife Trade Regulation, Article 8(3)(d).

[64] CITES. Resolution Conf. 12.10 (Rev. CoP15), "Registration of Operations that Breed Appendix-I Animal Species in Captivity for Commercial Purposes," p. 8.

abuses— blurring the distinction between animals born and bred in captivity on the one hand, and animals falsely labeled as such on the other. This has prompted a renewed focus on breeding operations—not out of concern for the animals' living conditions or welfare, but because of the regulatory challenges these operations present.

3.1.2 The Regulation of Trade in Captive-Bred Animals

To prevent fraud, CoP resolutions and EU regulations establish a detailed and technical framework for defining captive-bred animal specimens. In principle, the "breeding stock" must be maintained "without the introduction of specimens from the wild, except for the occasional addition of animals," for example, to prevent inbreeding.[65]

In principle, a specimen of a species listed in Appendix I (or Annex A under EU law) that was bred in captivity after being taken from the wild should not qualify for the exemption provided under Article VII(4)–(5). The commercialization of such a specimen should instead be subject to the standard dual-permit system, requiring both export and import permits.[66] In a recent judgment, the Court of Justice of the European Union clarified that the term "breeding stock" does not include the ancestors of specimens bred in a facility if those ancestors were never owned or held by that facility.[67] The Court further ruled that a specimen of an Annex A species held by a breeder could not be considered "born and bred in captivity" within the meaning of the regulation if its ancestors who are not part of the breeder's own breeding stock were acquired by a third party, prior to the regulation's entry into force, and in a way that was detrimental to the survival of the species in the wild.[68]

To combat fraud related to captive breeding, the CoP has sought to tighten the requirements for registering breeding facilities[69] and for

[65] CITES. Resolution Conf. 10.16 (Rev. CoP19), "Specimens of Animal Species Bred in Captivity"; Implementing regulation, Article 54.

[66] CITES. Resolution Conf. 11.16 (Rev. CoP15), "Ranching and Trade in Ranched Specimens of Species Transferred from Appendix I to Appendix II," second paragraph.

[67] Case C-659/20, *ET v Ministerstvo životního prostředí*, Judgment of 8 September 2022, ECLI:EU:C:2022:639 (Perroquets Ara hyacinthe).

[68] Ibid.

[69] CITES. Resolution Conf. 12.10 (Rev. CoP15), "Registration of Operations that Breed Appendix-I Animal Species in Captivity for Commercial Purposes."

marking animals.[70] More broadly, marking is considered a key component of the regulated trade system.[71] The CoP resolutions recommend a marking system that takes due account of "the humane care, well-being and natural behavior of the specimen concerned."[72] They encourage the use of implantable transponders provided this is "consistent with the well-being of the specimens concerned."[73] However, the resolutions fall short of defining what constitutes humane marking.

EU law is somewhat more ambitious. The heading of Article 67 of the Implementing Regulation is explicitly titled "Humane marking methods." According to this provision, the marking of live animals within the Union "shall be undertaken with due regard to humane care, well-being and natural behavior of the specimen concerned."[74] While this statement may appear aspirational, it is nonetheless a legal obligation. Article 66 of Regulation 865/2006 sets out the specific marking methods to be used within the Union: a seamlessly closed leg-ring for live birds, and a microchip transponder for other live vertebrates. However, competent national authorities may consider alternative methods, or exemptions from marking depending on the physical and behavioral properties of the animal.[75]

Ultimately, international and EU law protect the welfare of wild animals only insofar as they become the subject of commercial transactions—not beforehand, whether they have been taken from the wild or bred in captivity. Their welfare is considered only once the animal enters the trade system. From that point on, the way they are treated determines, at the very least, the legality of cross-border commerce.

3.2 *The Treatment of Animals as Merchandise*

The only provisions in CITES and the EU Wildlife Trade Regulation that explicitly address animal welfare concern the conditions of transport (3.2.1.) and housing at the destination (3.2.2.), and only after the animal

[70] CITES. Resolution Conf. 10.16 (Rev. CoP19), "Specimens of Animal Species Bred in Captivity."

[71] Article VII(7) of CITES; Implementing Regulation, sixth recital.

[72] CITES. Resolution Conf. 7.12 (Rev. CoP15), "Marking requirements for trade."

[73] CITES. Resolution Conf. 8.13 (Rev. CoP17), "Use of coded-microchip implants for marking live animals in trade."

[74] Commission Regulation (EC) No 865/2006 of 4 May 2006 laying down detailed rules concerning the implementation of Council Regulation (EC) No 338/97 on the Protection of Species of Wild Fauna and Flora by Regulating Trade Therein.

[75] Ibid., Article 66(2–3).

has been sold. While these requirements are primarily rooted in animal ethics, they are also connected to environmental ethics, insofar as "the improper transport of live specimens can increase the risk for injury or mortality, which may [result] in additional take of wild specimens."[76] Likewise, the aim of ensuring appropriate housing conditions is to support the long-term survival of the animal, thereby reducing the need to replace it in species conservation efforts.[77]

3.2.1 *Adequate Transport Conditions*

Under CITES, the issuance of export permits is systematically conditioned on evidence "that any living specimen will be so prepared and shipped as to minimize the risk of injury, damage to health or cruel treatment."[78] This concern extends across the entire journey—during transit, handling or transportation, and in particular during customs clearance or other formalities.[79] The EU Wildlife Trade Regulation also incorporates this requirement for export permits.[80] More broadly, it establishes this standard as a general obligation governing the movement of live specimens into, from, or within the Union.[81]

The Secretariat of the Convention is responsible for examining the "standards for appropriate preparation and shipment of living specimens."[82] In response, the CoP adopted a resolution specifically addressing the "transport of live specimens."[83] It encourages adherence to the "Live Animals Regulations" of the International Air Transport Association, as well as the 2013 CITES Guidelines for the Non-Air Transport of Live Animals and Plants (amended in 2022). The resolution further recommends that these detailed instruments be incorporated into national legislation and that the issuance of export permits be made conditional on compliance with these standards.

In the EU, transporters of live specimens must comply with Regulation 1/2005 on the protection of animals during transport and related

[76] CITES. Resolution Conf. 10.21 (Rev. CoP19), sixth paragraph.

[77] European Commission and TRAFFIC. 2020. *Reference Guide to European Union Wildlife Trade Regulations*. Belgium, Brussels, p. 101.

[78] Article III(2)(c), IV(2)(c), and V(2)(c) of CITES.

[79] Article VIII(3) of CITES.

[80] Article 5(2)(c) of the Wildlife Trade Regulation.

[81] Article 9 of the Wildlife Trade Regulation.

[82] Article XII(2) of CITES.

[83] CITES. Resolution Conf. 10.21 (Rev. CoP19).

operations.[84] However, in its proposal to revise Regulation 1/2005, the European Commission recommended excluding wild animals from the scope of the regulation.[85] Since its adoption is not expected before 2026 at the earliest, it remains too soon to determine the final content of the revised version. Ideally, the legislative process should be viewed as an opportunity for the EU to establish more specific standards for the transport of traded wild animals—potentially through an implementing regulation under the Wildlife Trade Regulation.

The various international and EU standards are designed to ensure a minimum level of animal welfare during transport. They place limits on journey duration, prohibit the transport of unfit animals, regulate transport methods and equipment, require appropriate training and qualifications for personnel, and mandate regular inspections, adequate space, and proper feeding. In addition, EU law requires that Member States sanction violations related to the inadequate preparation of live animal transport, particularly when such failures do not sufficiently minimize the risk of injury, disease, or cruel treatment.[86] Moreover, the European Commission may impose restrictions on the introduction into the Union of "live specimens of species listed in Annex B which have a high mortality rate during shipment or for which it has been established that they are unlikely to survive in captivity for a considerable proportion of their potential life span."[87]

3.2.2 *Appropriate Housing Standards*

Compared to provisions related to transport conditions, CITES provides shorter provisions on the conditions at the destination State, and only a limited number of specimens are subject to the relevant provisions. First, live specimens listed in Appendix I can only be granted an import permit if "the proposed recipient of a living specimen is suitably equipped to house and care for [them]."[88] The required evidence should therefore focus on the quality of the housing facilities, not on the animal's long-

[84] Article 9(5) of the Wildlife Trade Regulation.

[85] European Commission. 2023. Proposal for a Regulation of the European Parliament and of the Council on the Protection of Animals During Transport and Related Operations, amending Council Regulation (EC) No 1255/97 and repealing Council Regulation (EC) No 1/2005, COM/2023/770 final.

[86] Article 16 of the Wildlife Trade Regulation.

[87] Article 4(6)(c) of the Wildlife Trade Regulation.

[88] Articles III(3)(b) and III(5)(b).

term care. Second, live specimens of Appendix II species introduced from the sea are only allowed if they "will be so handled as to minimize the risk of injury, damage to health or cruel treatment."[89] In this case, the evidence should ideally speak to both the adequacy of the facilities and to the standard of care. Yet, no resolution has so far spelled out what exactly is meant by "suitably equipped facilities" or what constitutes "cruel treatment."

EU law provides a more comprehensive regulatory regime for housing requirements. Import permits for live specimens listed in both Annex A and Annex B[90] may only be granted if traders can demonstrate that appropriate housing is available at the destination and adequately equipped to conserve and care for the specimen properly.[91] In principle, the movement of live Annex A specimens within the EU is also conditional on the availability of suitable facilities at the destination.[92]

Within the Union, the holder of an Annex B specimen may only transfer it "after ensuring that the intended recipient is adequately informed of the accommodation, equipment and practices required to ensure the specimen will be properly cared for."[93] This requirement aims to encourage sellers of live animals to provide buyers with adequate guidance on how to house and care for them appropriately.[94]

Captive breeding establishments are considered appropriate facilities in the destination State under both CITES and EU regulations. In the case of zoos, circuses, and *traveling* menageries, international and EU law even relaxes the usual strict trade rules, recognizing the frequent cross-border movements of these mobile exhibitions. Traders of these specimens are exempt from standard export and import permits.[95] Instead, a single traveling exhibition certificate—valid for multiple uses—facilitates their movement.[96] To benefit from this exemption, however, the animals must either have been acquired before their species was listed under an Appendix or

[89] Article IV(6)(b) of CITES.

[90] Not just those introduced from the sea.

[91] Articles 4(1)(c) and 4(2)(b) of the Wildlife Trade Regulation.

[92] Article 9(2) of the Wildlife Trade Regulation.

[93] Article 9(4) of the Wildlife Trade Regulation.

[94] European Commission and TRAFFIC. 2020. *Reference Guide to European Union Wildlife Trade Regulations*. Belgium, Brussels, p. 101.

[95] Article VII(7) of CITES; Implementing Regulation, Article 31.

[96] Implementing Regulation, Article 30; CITES. Resolution Conf. 12.3 (Rev. CoP19), "Permits and Certificates," p. 10–12.

have been bred in captivity.[97] Therefore, traveling exhibitions involving animals that regularly cross borders are not, in themselves, regarded as inadequate facilities or as a form of harsh treatment. Only CITES adds a condition: the exemption applies only if there is proof that "any living specimen will be so transported and cared for as to minimize the risk of injury, damage to health or cruel treatment."[98] Notably, the EU regulation does not reiterate this requirement.

Fixed and stationary facilities—such as zoological parks—as well as nomadic ones that do not cross borders are likewise not considered inadequate facilities under international and EU law. However, specimens listed in Appendix I or Annex A may not be imported for primarily commercial purposes. In principle, this means they should not be acquired at the destination by commercial establishments whose primary aim is to generate profit.[99] However, commercial intent can often be obscured by invoking scientific or educational purposes linked to conservation. As a result, the conservation objective enables the profitable use of Annex A specimens within the Union, supporting activities such as captive breeding, research, and education.[100]

In conclusion, the legal framework established by CITES and its implementation within the EU reflects an environmental ethic that permits the trade of endangered species to satisfy both the basic needs of local populations and the desires of consumers—often with little concern for the animal's wild nature or the conditions in which captive animals are kept. As with any ethical dilemmas involving commerce, prohibition is dismissed as counterproductive, with the argument that it would only fuel illegal trade. Yet opting for a regulated trade effectively denies the intrinsic value of wild animals—their worth in and of themselves—and blocks any deeper questioning of the legitimacy and persistence of certain human practices, which are implicitly treated as quasi-essential.

[97] Article VII(7)(b) of CITES; Implementing Regulation, Article 30(1).

[98] Article VII(7)(c) of CITES.

[99] CITES. Resolution Conf. 5.10 (Rev. CoP19), "Definition of 'Primarily Commercial Purposes.'"

[100] Article 8(3) (f) and (g) of the Wildlife Trade Regulation.

Using Consumer Law to Improve the Treatment of Animals in Industrial Farming Production

Isabelle Doussan

1 Introduction

An increasing proportion of people in Western societies live in urban areas, and the way they view animals has changed considerably over the past century. Two phenomena in particular have led to such changing views: the disappearance of animal exploitation from public spaces, and the growing number of companion animals. The end of animal use for transport purposes in the beginning of the twentieth century led to the disappearance of horses, particularly in cities. This disappearance followed the removal of animals used for food from the public eye, starting in the nineteenth century, as slaughterhouses gradually moved outside urban areas, mainly for sanitary reasons. Before then, butchers and restaurant owners

I. Doussan (✉)
INRAE, GREDEG, CNRS Université Côte d'Azur, Nice, France
e-mail: isabelle.doussan@inrae.fr

A.-S. Epstein, A. Di Concetto (eds.), *Animals in EU Economic Law*, The Palgrave Macmillan Animal Ethics Series,
https://doi.org/10.1007/978-3-032-16171-0_10

often raised and slaughtered farmed animals near their businesses.[1] The industrialization of livestock production in the 1960s, and the consequent move of farming operations to the countryside, now means that poultry, pigs, cows, and other animals are confined to indoor facilities for their whole lives, away from public view.[2] At the same time, the number of companion animals in urban households has skyrocketed, generating a flourishing market.

These societal changes have been reflected in academic research, with the rise of *animal studies*[3] and the growth of scientific knowledge about animal behavior. Humanities scholars have described this new way of approaching animals in society as constitutive of a "pathocentric turn,"[4] whereby society has adopted a benevolent attitude toward animals. As a result, society considers the pain inflicted on animals to be a form of injustice. The animal advocacy movement has thus become a social justice movement.[5]

These developments have been conducive to greater consideration from the general population regarding the treatment of animals used by humans, particularly farmed animals raised for food. The concerns of animal advocates are additionally supported by scientific data, which shows that industrial farm animal production, including transport and slaughter, generates a great deal of suffering and stress in animals. Historically, farm animal welfare rules have done little to limit animal suffering, as such rules are typically consistent with industry goals and have mitigated animal cruelty only to the extent that is economically feasible for producers.

This observation about animal welfare rules leads to a double hypothesis. The first is that legislatures in Western jurisdictions have enacted animal welfare standards as a compromise between the economic interests of industrial farm animal producers and legal requirements to recognize animal sentience. However, this attempt to find a mutually agreeable

[1] Rémy, C. 2009. *La fin des bêtes. Une ethnographie de la mise à mort des animaux*, 210. Paris: Economica.

[2] Baldin, D. 2016. Animaux à aimer, animaux à tuer. *Revue d'histoire des sciences humaines* 28: 40.

[3] Michalon, J. 2017. Les Animal Studies peuvent-elles nous aider à penser l'émergence des épistémès réparatrices? *Revue d'anthropologie des connaissances* 11(3): 321–349.

[4] Stépanoff, C. 2021. *L'animal et la mort. Chasses, modernité et crise du sauvage*, 384. Paris: La Découverte.

[5] Carrié, F., Doré, A., and J. Michalon. 2023. *Sociologie de la cause animale*, 126. Paris: La Découverte.

compromise is doomed to fail because industrial farm animal production systems are inherently incompatible with ensuring that animals live and die in conditions that meet their physical and psychological needs. A second hypothesis is that the social acceptability of these production methods relies in part on the hidden nature of the cruel and violent treatment to which factory-farmed animals are subjected, in a way that shields consumers from learning about or viewing such treatment. The fact that undercover videos and exposés, particularly on social media, have been the main drivers of recent progress for farm animal welfare aligns with such a hypothesis.[6] Consumer law, by upholding consumers' right to information, therefore has the potential to serve as a powerful tool in eliminating the cruelest production methods.

2 Animal Welfare and Industrial Productivity: An Impossible Compromise

2.1 *The Evolving Purposes of Animal Protection Laws*

French law has recognized animal sentience in several instances, in laws dating back as early as the mid-nineteenth century. However, throughout history, the recognition of animal sentience did not always serve the same goals. The French legislature recognized animal sentience for the first time in 1850, with the enactment of the Grammont Law, which punished "those who publicly and abusively mistreat domestic animals" with prison sentences and fines.[7] For the first time in French legal history, the Grammont Law limited the rights humans could exercise over an animal.[8] More specifically, even though French law has qualified animals as property,

[6] For instance, see the description of the campaign actions of L214, a French animal protection organization, in the parliamentary work on the Law n° 2018–938 of October 2018 and Law n° 2021–1539 of 30 November 2021. For more on the "politics of sight" and animal abuse, see Pachirat, T. 2011. *Every Twelve Seconds: Industrialized Slaughter and the Politics of Sight.* New Haven: Yale University Press.

[7] Loi du 2 juillet 1850 sur les mauvais traitements envers les animaux domestiques, Moniteur Universel, 3 juill. 1850, p. 2269. The law consists of a single article which states that "those who publicly and abusively mistreat domestic animals will be punished by a fine of 5 to 15 francs and may be imprisoned for one to five days. The prison sentence will always be applied in the event of a repeat offence."

[8] As per Article 544 of the French Civil Code, which states that "Ownership is the right to enjoy and dispose of things in the most absolute manner, in accordance with laws or regulations."

and has allowed anyone to destroy their own property, the Grammont Law determined for the first time that abuse of a domestic animal was not an attribute of property rights, additionally criminalizing such acts.

However, the Grammont Law was limited in its ability to protect animals in two important ways. First, the prohibition on abusing animals only applied to mistreatment carried out "publicly and abusively." The Legislature's main objective in the nineteenth century was thus to punish the public exhibition of people abusing living beings, and the main objective of the Grammont Law was not to protect the animals for their own sake but rather to uphold a moral and social order. It aimed to prevent the public exhibition of violence and cruelty toward sentient beings, which society regarded as cruel and deviant behavior. This approach to regulating behavior was then underpinned by the idea that deterring "minor" violence against animals would "help curb major violence between humans."[9] A more general objective was also to educate the working class, as social workers at the time considered that "the coachman who abuses his horse is violent towards his family and, more broadly, to his social relations; whereas the one who drives his carriage gently, without brutality and without abusing the whip is more likely to be a good father and a submissive worker."[10]

The objective pursued by the French Legislature when enacting animal welfare legislation shifted under the effect of animal agriculture industrialization. As industrial farm animal production systems became dominant in France, the Legislature recognized the ability for animals to suffer (*sensibilité*). The French 1976 Nature Protection Act thus provided that farmed "animals are capable of suffering and so must be placed by its owner in conditions compatible with the biological needs of its species."[11] The recognition of animal sentience became further codified in the Treaty on European Union in 1997, and later in 2007, in the Treaty on the Functioning of the European Union (TFEU), which states that "the Union and the Member States shall take full account of the welfare requirements of animals as sentient beings [...] when formulating and implementing the Union's policies on agriculture, fisheries, transport, the

[9] Translated from Agulhon, M. 1981. Le sang des bêtes. Le problème de la protection des animaux en France au XIXème siècle. *Romantisme* 31:81–110.

[10] Pierre, E. 2007. Réformer les relations entre les hommes et les animaux: fonction et usages de la loi Grammont en France (1850–1914). *Déviance et Société* 1 31: 65–76.

[11] Code rural et de la pêche maritime, art L. 214–1. The wording has stayed the same since 1976.

internal market, research and technological development and space."[12] In 2015, the French legislature amended the Civil Code to include a general recognition of animals' capacity to feel. Article 515–14 now explicitly states: "Animals are living beings capable of suffering."

The emergence of new threats has often led to the creation of new values and legal concepts. Following this pattern, the recognition of animals' ability to feel as a new protected interest can be interpreted as a legal response to the transformation of animal agriculture,[13] as legislatures and courts must deal with the institutional and systemic violence of industrial farm animal production, which became dominant in Western jurisdictions in the second half of the twentieth century. In this context, legislatures today pursue different objectives than the French legislature of the nineteenth century, which was mostly concerned with deterring morally reprehensible behavior.[14] In the context of factory farming, the violence visited upon animals does not result from isolated individuals misbehaving. Instead, this type of violence is systemic and inherent to industrial farm animal production systems.

The purpose of legislatures in the West is thus to find a compromise between the conflicting interests of producers on the one hand, and the interests of animals not to suffer on the other. Legislatures thus distinguish between acts that are detrimental to animals (in other words, acts that are likely to generate suffering, stress, and discomfort) but are legal because they are necessary, and those detrimental acts that are determined to be unnecessary and thus illegal. For instance, the French legislature considers it a small offense to deliberately kill an animal when such killing is not necessary and outside the scope of regulated business activities.[15]

[12] Article 13 of the TFEU provides a broad-scope exemption covering "the legislative or administrative provisions and customs of the Member States relating in particular to religious rites, cultural traditions and regional heritage." The French version of the text only recognizes the ability of animals to suffer ("sensibilité"), as opposed to sentience in the English and other linguistic versions. See Di Concetto, A. 2023. L'animal dans la ressource. In *La prise en compte par le droit des enjeux de l'alimentation durable*, ed. Fanny Giansetto. Paris: Mare & Martin (in French).

[13] Porcher, J. 2014. *Vivre avec les animaux, une utopie pour le XXIe siècle*. Paris: La Découverte.

[14] For more on this, see Di Concetto, A. 2023. La protection de l'animal par le droit pénal: enjeux éthiques. In *Le droit pénal face à l'éthique*, ed. Julie Gallois and Chloé Liévaux. Paris: Dalloz (in French).

[15] Code rural et de la pêche maritime, art 522–1. Also see Di Concetto, A. 2023. La protection de l'animal par le droit pénal: enjeux éthiques. In Le droit pénal face à l'éthique, ed.

Similarly, if an economic operator abuses or allows animals to be abused, this offense would only be punishable if the abuse was unnecessary,[16] and certain practices that are particularly harmful to animals are only allowed in cases of "absolute necessity."[17] In addition to these prohibitions and obligations not to inflict "unnecessary" suffering, French law mandates certain animal protection measures, such as the obligation to stun animals before slaughter, the obligation to comply with specifications related to the design and maintenance of farming buildings, and the ability to allow certain animals outdoor access.[18]

2.2 *The Evolving Content of Animal Welfare Standards*

The content of animal welfare standards in the Global North is shaped and informed by discussions and compromises between three main actors: producers, scientists in animal welfare science and ethology, and society.[19] A recent example of discussions between these three groups resulting in new law is the inclusion of cephalopods and fetal forms of mammals in the scope of EU Directive 2010/63 on the Protection of Animals Used for Scientific Purposes, which was adopted on the basis of scientific data demonstrating these animals' ability to experience suffering and distress.[20]

Julie Gallois and Chloé Liévaux. Paris: Dalloz (in French); Leborne, J. 2012. Le droit pénal du bien-être animal. *Revue Semestrielle de Droit Animalier* 2/2012 (in French).

[16] Code rural et de la pêche maritime, art L215–11.

[17] Examples of these practices include the use of confinement devices, such as fences, cages or more generally, any form of keeping that is inadequate to meet a certain species' needs, or likely to cause injury or suffering.

[18] For example, the Council of Europe, European Convention for the Protection of Animals During Transport, 6 November 2003, E.T.S. 193. The Council of Europe Conventions forms the basis for EU farm animal welfare legislation such as Council Directive 1999/74 of 19 July 1999 Laying Down Minimum Standards for the Protection of Laying Hens, 1999 O.J. (L 203) 53–57 and Council Directive 2008/119 of 18 December 2008 Laying Down Minimum Standards for the Protection of Calves, 2009 O.J. (L 10) 7–13. For a history of EU farm animal welfare legislation, see Di Concetto, A. 2022. The Double Edged Sword: International Law and Its Effects on EU Farm Animal Welfare Legislation. *Global Journal of Animal Law, Special Issue: Internal Law and Animal Health and Protection: Persistent Themes, New Prospects for Change* 10(2).

[19] Bismuth, R., Demaret, A., Di Concetto, A., Epstein, A., Rouxel, M., and Y. Soubigou. 2018. La concurrence des normativités au cœur de la labellisation du bien-être animal. *Revue internationale de droit économique* XXXII: 369–393 (in French).

[20] Directive 2010/63/EU of the European Parliament and of the Council of 22 September 2010 on the protection of animals used for scientific purposes, 2010 O.J. (L 276) 33–79.

When it comes to farmed animals, European law often explicitly states that texts relating to animal welfare must be based on scientific data concerning their physiological and ethological needs, and that such texts must be regularly updated to reflect scientific progress. For instance, the European Convention for the Protection of Animals Kept for Farming Purposes provides for the creation of a Standing Committee, which is to be responsible for monitoring the "development in scientific research and new methods in animal husbandry"[21] and for drawing up recommendations for its implementation, "based on scientific knowledge concerning the various species of animals."[22] Another example of how the EU attempts to align its animal welfare provisions with current science is Regulation 1099/2009 on the Protection of Animals at the Time of Killing, which contains numerous references to scientific and technical knowledge.[23] The European Commission is also required to designate animal welfare reference centers to support the actions of the Commission and Member States.[24] These centers must provide the best available scientific and technical research, particularly in animal behavior, psychology, genetics, health, and nutrition. The role of these centers is also to create indicators to improve animal welfare, and to analyze the welfare of animals used for commercial and scientific purposes.

A challenge for legislatures has been to incorporate ever-changing knowledge about animals, given that animal welfare science has continually evolved over the past 20 years.[25] In the process, the very definition of "animal welfare" has also changed. The field of animal welfare science first emerged as a sub-field of zootechnics and was subsequently shaped by

[21] Council of Europe, European Convention for the Protection of Animals kept for Farming Purposes, art 9(2), 10 March 1978, E.T.S 87.

[22] Ibid., art 9(1).

[23] For instance, Council Regulation 1099/2009 of 24 September 2009 on the Protection of Animals at the Time of Killing, art 4(2), 15(4), and 13(2), 2009 O.J. (L 303) 1–30.

[24] Regulation 2017/625 on official controls and other official activities performed to ensure the application of food and feed law, rules on animal health and welfare, plant health and plant protection products, art 95, 2017 O.J. (L 95/75) 1–142. See also Bouhier, V. 2021. Le bien-être animal et le droit primaire dérivé: une exigence perfectible. *Revue de l'Union européenne* 9: 454–460.

[25] For example, on pain in farm animals, see Sabbagh, C. and P. Le Neindre. 2013. *Douleurs animales en élevage*. Versailles: éditions Quae, p. 128 (Matière à débattre et décider series) (in French); and on animal consciousness in Le Neindre, P., Dunier M., Larrère R., and P. Prunet (coord.). 2018. *La conscience des animaux*. Versailles: éditions Quae, p. 118. (coll. Matière à débattre et décider) (in French).

various other scientific disciplines. In the early days of factory farming, animal welfare consisted of attempting to adapt animals to industrial environments typical of factory farming, such as those with high animal density, accelerated biological cycles, and exposure to stress. Scientists at the time did not intend to adapt the farming environment to animals—unlike for plants[26]—and only considered animal welfare as a component of animal health.[27] The dominant view among zootechnic scientists was that animal welfare was a technical and medical concept in support of economic rationality. Scientific disciplines other than zootechnics later contributed to shaping animal welfare science. An example of such a discipline would be ethology, which focuses on animals' subjective experiences. Ethologists have demonstrated that, in addition to pain sensitivity, animals interact with their environment and have different levels of consciousness depending on the species.

As animal welfare science has evolved, so has the notion of animal welfare, an amorphous concept open to a wide variety of interpretations.[28] For example, in 2018, the French food safety administration (*Agence nationale de sécurité sanitaire de l'alimentation, de l'environnement et du travail*, ANSES), adopted the following definition of animal welfare: "…the welfare of an animal is the positive mental and physical state in relation to the satisfaction of its physiological and behavioral needs, as well as its expectations. This state varies according to the animal's perception of the situation."[29] This new definition, which informs ANSES's research mandate on animal welfare topics, thus takes into account more modern views of animal welfare science, and animals' ability to have negative and positive experiences. This new definition also shifts the perspective of scientists, by mandating that they should adopt the animal's point of view. Animal welfare indicators based on the animals themselves, such as the

[26] See, for example, Rosier, B. and J. Berlan. 1989. Les nouvelles technologies agricoles comme production sociale. *Économie rurale* 192–193: 23–28.

[27] Dupin, L. 2021. Bien-être animal et avancées scientifiques. *Revue Semestrielle de Droit Animalier* 2: 597–622.

[28] Desmoulin-Canselier, S. 2021. Des difficultés révélatrices d'une véritable aporie? *Revue de l'Union européenne* 652: 567–579. Dupin, L. 2021. Animal welfare and scientific advances. *RSDA* 2: 597–622.

[29] Anses. 2018. *Avis relatif au "Bien-être animal: contexte, définition et évaluation."* 16 February 2018, p. 16 (in French). Available at:https://www.anses.fr/fr/content/l%E2%80%99anses-propose-une-d%C3%A9finition-du-bien-%C3%AAtre-animal-et-d%C3%A9finit-le-socle-de-ses-travaux-de.

Welfare Quality© project,[30] have also contributed to the shift to a more modern approach to animal welfare science.

Despite these considerable developments in the science of animal welfare, farm animal welfare standards in EU law have not been revised to guarantee the protection of animals' physiological needs. Instead, animal welfare laws still authorize a host of harmful practices.[31] The discrepancy between these rules and scientific recommendations is so significant[32] that it calls into question the use of the term "animal welfare" when referring to these laws.[33]

This situation has led to increasing societal demand for improving the treatment of farm animals, especially during transport and slaughter.[34] Yet, it seems that industrial farm animal production methods can no longer meet animal welfare requirements as scientifically defined today. Members of the scientific community have also come to the realization that their professional contributions seem to have produced little tangible effect on the treatment of farmed animals, and that their discipline has even "served as a scientific guarantee for production methods that generate serious violations of animal welfare," leading them to explicitly state that "improving animal welfare in intensive systems is not possible."[35]

[30] Veissier, I., Botreau, R., and P. Perny. 2010. Évaluation multicritère appliquée au bien-être des animaux en ferme ou à l'abattoir: difficultés et solutions du projet Welfare Quality. *Inra Prod. Anim.* 23(3): 269–284; Mounier, L. (coord.). 2021. *Le bien-être des animaux d'élevage. Évaluer le bien-être animal.* Versailles: éditions Quae, p. 75. (coll. Les mémos) (in French).

[31] Del Cont, C. 2021. L'animal d'élevage dans le droit de l'Union européenne ou la difficile conciliation de l'objectif de bien-être animal et des objectifs économiques. *Revue de l'Union européenne* 652: 546–553.

[32] See, for example, Broom, D.M. 2018. Legally accepted pain and other animal welfare issues, in Hild S., Schweitzer L. (dir.), *Le bien-être animal: de la science au droit.* Paris: L'Harmattan, 299–316. See, for example, European Court of Auditors. 2018. *Animal welfare in the EU: bridging the gap between ambitious targets and reality of implementation.* Special Report 18. Available at: https://www.eca.europa.eu/Lists/ECADocuments/SR18_31/SR_ANIMAL_WELFARE_FR.pdf.

[33] Mounier, L. (coord.). 2021. Le bien-être des animaux d'élevage. Évaluer le bien-être animal. Versailles: éditions Quae, p. 15. (coll. Les mémos) (in French).

[34] See in particular, Magdelaine, P., Roguet, C., Dockès, A.-C., and E. Delanoue. 2018. The livestock controversy. Principaux enseignements du projet ACCEPT. *Revue TeMA* 47. Available at: http://accept.ifip.asso.fr/docs/publication/magdelaine_tema_n_47_vf.pdf.

[35] We translated Leterrier, C., Aubin-Houzelstein, G., Boissy, A., Deiss, V., Fillon, V., Lévy, F., Merlot, E., and O. Petit. 2022. Améliorer le bien-être des animaux d'élevage: est-ce toujours possible. Revue Sésame, 28 June 2022. Available at: https://revue-sesame-inrae.fr/ameliorer-le-bien-etre-des-animaux-delevage-est-ce-toujours-possible/consulté on 19/02/2024.

2.3 The Significant Obstacles to Legal Reform

Closing the gap between animal welfare science and animal welfare standards in the law would likely require increasing production costs of animal-based food.[36] Before even considering that step though, such an effort would also require a strong critique of industrial farm animal production within legislative and regulatory procedures. However, animal welfare legislation is shaped by procedures that seek compromise between stakeholders, leaving little to no room to challenge factory farming.

When it comes to farm animal welfare, the legislative and rule-making process—especially at EU-level—is not designed to take animal sentience into account. This is due to several factors, starting with the fact that the industrial farm animal industry is well represented to the competent authorities and in consultative working groups. Animal protection groups, on the other hand, face challenges when trying to make their voices heard by the EU institutions. For example, the requirements that organizations must meet to be represented in policy discussions in the EU institutions hinder fair and equitable representation of non-profit groups. These requirements also tend to exclude the most vocal groups from rule-making processes.[37]

This situation largely explains why the reform announced by the European Commission in its "Farm-to-Fork Strategy" has stalled,[38] starting with the revision of EU farm animal welfare legislation to "includ[e] … animal transport and the slaughter of animals, to align it with the latest scientific evidence, broaden its scope, make it easier to enforce and ultimately ensure a higher level of animal welfare."[39] By the end of its term

[36] At least if the negative externalities of factory farming, particularly on the environment, are not internalized. Ducourant, S. 2021. Science or Ignorance of Animal Welfare? A Case Study: Scientific Reports Published in Preparation for the First European Directive on Animal Welfare (1979–1980). *Science, Technology, & Human Values*:1–28.

[37] "The expertise required to decipher official, legal, and political mandates is a major obstacle for public interest groups, which have limited means and are therefore far less able to attract EU public affairs specialists compared to the private sector. Furthermore, over the years, English has become the main working language of the European institutions, especially in the European Commission." We translated Di Concetto, A. 2023. Le pouvoir des lobbies privés face au mouvement de protection animale. In *Les mutations contemporaines du droit de l'animal*, ed. O. Le Bot, 139–156. Aix-en-Provence: DICE Éditions.

[38] European Commission. 2020. A Farm-to-Fork Strategy for a Fair, Healthy and Environmentally-Friendly Food System. COM/2020/381 final.

[39] Ibid., p.8.

in 2024 though, the European Commission had only published one of the four proposals for new legislation on farm animal welfare—the Proposal for a new regulation on the transport of animals.[40] The European Commission announced that the legal reforms to end the use of cages in animal agriculture would be delayed until 2026, but it provided no specific timeline for how the process for such reforms would proceed, instead insisting on the need to establish a dialogue with farmers.[41]

Given the roadblocks animal advocates now face in the Legislature, a potentially effective solution to make the rule-making process fairer would be to reinforce the public's right to information about the practices used in industrial farm animal production. One way to achieve this is by substantiating consumers' right to make an informed choice between various animal-based products.

3 Consumers' Right to Information

3.1 The Increasing Criminalization of Investigations on Factory Farms

Since the 1850 Grammont Law, the French legislature has pursued the objective of prohibiting public acts of violence toward animals. However, the interpretation of these laws has sometimes penalized those who expose abuse rather than those who commit it. For instance, in France, recording footage of animal abuse constitutes an act of complicity, punishable by the same penalties as those incurred by the perpetrator. The online dissemination of these images is also a specific offense in French criminal law.[42] The law further provides that the presence of an underage individual increases the severity of animal cruelty offenses.[43]

[40] Proposal for a Regulation of the European Parliament and of the Council on the Protection of Animals During Transport and Related Operations, Amending Council Regulation (EC) No 1255/97 and Repealing Council Regulation (EC) No 1/2005. Available at: https://food.ec.europa.eu/system/files/2023-12/aw_in-transit_reg-proposal_2023-770_0.pdf.

[41] European Commission. 2024. Final Report of the Strategic Dialogue on the Future of EU Agriculture. 4 September 2024, p.66. Available at: https://agriculture.ec.europa.eu/document/download/171329ff-0f50-4fa5-946f-aea11032172e_en?filename=strategic-dialogue-report-2024_en.pdf.

[42] Code pénal, art 521-1-2. This provision covers domestic, wild, and captive animals.

[43] Code pénal, art 521-1. Similarly, Article 521-1-1 of Code pénal criminalizes sexual abuse on an animal and increases penalties when such an offense is committed in the presence of a minor.

Similarly, the protection of the economic profitability of industrial farm animal production systems is still an objective pursued by legislatures in Western jurisdictions. So-called Ag Gag Laws, which have proliferated in the United States and Canada, would be an example of measures meant to maximize profits over revealing evidence of wrongdoing. These laws criminalize whistleblowers for investigating, documenting (including by recording footage), and publicizing evidence of animal abuse and malfeasance taking place on factory farms.[44] France has adopted a similar law restricting the ability of animal advocates to disclose the treatment of animals on factory farms.[45] In support of the largest French farmers' union (the *Fédération nationale des syndicats d'exploitants agricoles*, FNSEA), the French Legislature enacted a law in 2019, which created a special police enforcement unit to fight anti-farming speech ("agribashing"). In addition to being tasked with investigating and prosecuting trespass on farms, this unit also conducted a political mission targeting environmental and animal advocates. In a 2022 ruling, the Paris Administrative Court underscored that the purpose of this special unit was to prevent criminal behavior, as well as "actions of an ideological nature" that could consist of "symbolic denigrating actions," including "trespass on farms or agri-food establishments for the purpose of carrying out symbolic actions," "anti-fur actions," and "actions by certain anti-speciesist groups against hunting, an activity which is related to farming."

The Paris Administrative Court subsequently ruled that the French government had overstepped its powers in ordering the police to investigate "actions of an ideological nature,"[46] but this decision was reversed on appeal by the French Supreme Administrative Court in 2024.[47] Such repressive and censorious initiatives by governments reinforce the culture of secrecy and lack of transparency in industrial farm animal production,[48]

[44] See Fiorentino, A. 2016. Les lois "Ag Gag" américaines: l'interdiction de filmer la mort de l'animal. *Revue Semestrielle de Droit Animalier* 1:399–410 (in French); Marceau, J.F. 2015. Ag Gag Past, Present, and Future. *Seattle University Law Review* 38(4); Lazare, J. 2022. Animal Rights Activism and the Constitution: Are Ag-Gag Laws Justifiable Limits? *Osgoode Hall Law Journal* 59(3).

[45] More generally, any agricultural practices.

[46] TA Paris, 1er févr. 2022, n° 2006530, 2018140/3-1.

[47] Conseil d'État, 7 nov. 2024, n° 488664.

[48] For more on the opacity of slaughterhouses and the invisibility of the treatment of animals at slaughter, see Falorni, O. and J.-Y. Caullert. 2016. Rapport fait au nom de la commission d'enquête sur les conditions d'abattage des animaux de boucherie dans les abattoirs français. *Assemblée nationale* 4038:316. See also Sinclair, U. 1906. *The Jungle*. United States:

but these laws also reveal the significant impact of actions that publicly disclose the ways in which farm animals are treated on factory farms.

3.2 *The Right of Citizens to Learn the Findings From Undercover Investigations*

Because animals' ability to feel pain is a legally protected interest, some legal courses of action are available to animal advocates. First, information related to the treatment of animals is considered a general interest and so, in French law, the publication of images displaying animal abuse is not a criminal offense "when the recording, possession, dissemination, or consultation of these images is intended to contribute to a public debate of general interest, or to serve as evidence in court."[49] A similar precedent exists in international law. In 1999, a ruling by the European Court of Human Rights (ECHR) determined that the publication of an article on the cruel methods of seal hunting in Norway was consistent with "the vital public interest in ensuring an informed public debate over a matter of local and national as well as international interest."[50] The Grand Chamber of the ECHR also ruled specifically on the issue of farm animals, in a case dealing with a TV commercial alternatively showing pigs confined on a factory farm and living in the wild, and displaying messages encouraging viewers to eat less meat. The Court ruled that the commercial was "of definite public interest" in the sense of Article 10(2) of the Convention, since "it concerned consumers' health and the protection of animals and the environment."[51]

Another example of international law upholding freedom of information regarding the treatment of animals used for food can be found in French case law. In a 2022 ruling, the *Cour de Cassation* (France's highest criminal court) overturned a lower court's ruling for failing to properly

Doubleday, Jabber & Company; Epstein, A.-S. and T. Perroud. 2022. Lanceurs d'alerte: l'industrie des productions animales est plus opaque que celle du nucléaire. Le Monde, 18 October.

[49] Article 521–1-2 of the French Penal Code.

[50] European Court of Human Rights. 1999. *Bladet Tromsø and Stensaas v. Norway*, no. 21980/93, 20 May 1999, § 73.

[51] European Court of Human Rights (Grand Chamber). 2009. *Verein gegen Tierfabriken Schweiz v. Switzerland*, no. 32772/02, 30 June 2009, § 92. Nefussy-Venta, N. 2023. "N'est plus monarque dans sa maison celui qui porte atteinte à la cause animale!" *Les Petites Affiches* 2023, no. 5.

balance the interest of the general public to be informed about rabbit factory farms against the interest of a rabbit farmer, whose private property had been trespassed upon by members of L214, an animal protection organization, to record and publicize the treatment of rabbits. The French Court drew upon the case law of the ECHR when reaching such a conclusion, stating that: "…in accordance with the case law of the European Court of Human Rights, between two conventionally protected rights, the national court must always weigh up the interests involved in order to seek a balance between the competing rights and, where appropriate, give preference to the solution which best protects the most legitimate interest."[52]

3.3 *The Right of Consumers to Be Informed Through Food Labels*

Measures to improve transparency on factory farms can also be found in French consumer law, which lists consumer information "about the places and methods of production and processing of agricultural and agri-food products" as one of the objectives of the agricultural policy.[53] Additionally, EU law considers animal welfare to be a "societal demand" that must be taken into account through market forces and competition.[54] This strong societal demand has been further illustrated by an overwhelming majority of respondents to the public consultation (90%) who expressed their support for better consumer information, specifically through food labeling that assesses the treatment of farm animals, including at the transport and

[52] We translated Cour de cassation, 1ère chambre civile. 8 February 2023. Decision no. 22–10.542. Also see Verlhac, J. 2023. Droit associatif animalier. *Revue Semestrielle de Droit Animalier* 1:187–194. For a critical opinion, see Dassieux, N. 2023. Balancez, Balancez. *Recueil Dalloz*: p. 545.

[53] Article L 1-I 10° of French Code rural et de la pêche maritime.

[54] Regulation 2021/2115 Establishing Rules on Support for Strategic Plans to Be Drawn Up By Member States Under the Common Agricultural Policy, art 6(1)(i), 2021 O.J. (L 435/28). Also see Bismuth, R., Demaret, A., Di Concetto, A., Epstein, A., Rouxel, M., and Y. Soubigou. 2018. La concurrence des normativités au cœur de la labellisation du bien-être animal. Revue internationale de droit économique XXXII: 369–393 (in French); Di Concetto, A. 2023. Le bien-être animal dans la nouvelle Politique Agricole Commune. *Revue de droit rural* n°1/2023 (in French). Also see Chap. 3, Nicolas de Sadeleer and Alice Di Concetto, "The Common Agricultural Policy and Animals: Friends or Foes?"

slaughtering phases.[55] It is true that under certain conditions, labeling can be an effective tool to make the living and slaughter conditions of animals more visible.

There are two types of consumer information that a front-of-pack food label can present. The first type of information is "best practices," which have the effect of segmenting the market. The EU organic label is an example of this type of information. The second type of information is that which guides consumer choice, in a way that provides consumers with the ability to compare products. Information delivered by this second type of labels can also have strong effects on production methods—sometimes strong enough to prompt regulatory action.[56]

Industrial farm animal producers generally accept the idea of a label that promotes best practices but oppose labels that steer consumer choice toward more humane products. This position was clearly stated in the 2020 report by the French National Food Council (*Conseil national de l'alimentation*),[57] which aimed to inform the French public administration regarding the adoption of an animal welfare label on animal-based food products.[58] While industry groups were in favor of a label that would display best practices, animal protection, consumer protection, and environmental protection organizations, along with the national veterinary association, were all in favor of a food labeling system that would inform consumers about all production methods. Advocacy groups further recommended that this label should rely on science-based specifica-

[55] Ninety percent of respondents (53,128 out of 59,281) believed that an EU animal welfare label would be a useful tool for informing consumers about the treatment of farmed animals. Eighty-three percent of respondents (49,212 out of 59,281) expressed that an animal welfare label should be based on a variety of specifications, including transport and slaughter. Source: European Commission. 2022. Factual summary report of the online public consultation

in support of the fitness check and revision of the EU animal welfare legislation. Ref. Ares(2022)2359311, p.5. Available at: https://ec.europa.eu/info/law/better-regulation/have-your-say/initiatives/12950-Bien-etre-animal-Revision-de-la-legislation-de-l%E2%80%99Union/public-consultation_en.

[56] Muller, L. 2019. Nutri-score: quand l'étiquetage des aliments devient prescriptif. The Conversation, 26 August. See also Di Concetto, A. 2024. Farm Animal Welfare and Food Information for European Union Consumers: Harmonising the Regulatory Framework for More Policy Coherence. *European Journal of Risk Regulation* 15(1): 122–136.

[57] Conseil national de l'alimentation. 2020. Réflexion pour une expérimentation d'un étiquetage du mode d'élevage, Avis 85. Available at: https://cna-alimentation.fr/wp-content/uploads/2020/07/CNA_Avis85_Etiquetage-mode-%C3%A9levage.pdf.

[58] Ibid., p. 19.

tions.[59] Conversely, representatives for industrial farm animal producers cautioned that this type of label "presents the risk of establishing a hierarchy between farming methods."[60]

There is, in fact, a precedent for a label that informs consumers of the method of production of animal-based products, applicable to shell eggs on the EU market.[61] More specifically, EU law requires that all shell eggs sold on the EU market must be marked with a numerical code indicating the farming method used to keep egg-laying hens. These farming methods are cages, barn eggs, free-range eggs, and organic eggs.[62] Even though the primary goal of this measure has been to increase consumer information, animal law scholars have argued that "the regulation on European marketing standards for eggs is the piece of legislation that has done the most to promote the welfare of laying hens" and "has [...] been far more effective than legislation on animal welfare, which authorizes cage farming."[63] Further demonstrating that the method-of-production labeling on eggs has been effective in turning the egg industry away from using cages, caged farms primarily supply the food industry with egg products (such as liquid eggs), for which labeling is not mandatory.[64]

The egg labeling system indicates the conditions that a food label must meet to effectively inform consumers about the treatment of farm animals.[65] Firstly, the specifications used on the label must be designed by stakeholders, including animal protection and consumer organizations, producers, scientific experts, and public authorities.[66] Secondly, the label

[59] For details of the working group members' positions, see ibid., p. 29.

[60] Ibid., Appendix no. 9, p. 90.

[61] Commission Delegated Regulation 2023/2465 Supplementing Regulation 1308/2013 as Regards Marketing Standards for Eggs, art 11(2), 2023 O.J. (L 2023/2465).

[62] Ibid.

[63] Di Concetto, A. and M. Friant-Perrot. 2021. Le bien-être animal et l'information des consommateurs. *Revue de l'Union européenne* 9: 481–495.

[64] European Commission. 2019. *Evaluation of Marketing Standards Contained in the CMO Regulation*, November 2019, p.68. This report is cited by Di Concetto, A. and M. Friant-Perrot. 2021. Le bien-être animal et l'information des consommateurs. *Revue de l'Union européenne* 9: 481–495.

[65] Di Concetto, A. and A.-S. Epstein. 2023. EU Consumer Information as a Tool to Regulate the Treatment of Farm Animals. *EU Journal of Consumer Law*: 145–174. See also Degois, T. 2020. *La protection du bien-être animal au sein de l'Union européenne*. Rapport d'information, Commission des affaires européennes, Assemblée Nationale.

[66] See, for example, the ANSES Expert Committee on Animal Health and Welfare. Available at https://www.anses.fr/fr/content/sant%C3%A9-et-bien-%C3%AAtre-des-animaux-0. Accessed 15 May 2024.

should cover practices of all the economic actors involved in the production value chain, so as to inform consumers about the treatment of animals over their entire lives, such as the use of breeds, breeding practices, transport times, and slaughter conditions.[67] However, this food label should be easy to read for consumers. A good practice from that perspective is the Nutri-Score multi-level model, which summarizes a large amount of information in a digestible way.

Lastly, whether the label is mandatory or voluntary plays a central role in its effectiveness. Mandatory labels appear on all food products,[68] and are therefore more effective compared to voluntary labels, but public authorities tend not to be in favor of mandatory labels. However, reliable voluntary labels based on a large number of specifications already exist in certain Member States and could serve as a model for other Member States in the EU and at EU level. For instance, *Beter Leven* in the Netherlands[69] and *Étiquette bien-être animal* in France are examples of useful voluntary labels.[70] One key limitation of voluntary labels is that they tend to proliferate,[71] in a way that might cause confusion and distrust in consumers.[72]

[67] The information could be, for example, the use of slaughter methods such as electric bath for poultry or carbon dioxide gassing for pigs, which are lawful killing methods but not recommended by the European Food Safety Authority (EFSA). Di Concetto, A. and M. Friant-Perrot. 2021. Le bien-être animal et l'information des consommateurs. *Revue de l'Union européenne* 9: 481–495.

[68] For a favorable position on mandatory labeling, see Degois, T. 2020. *La protection du bien-être animal au sein de l'Union européenne*. Rapport d'information, Commission des affaires européennes, Assemblée Nationale, 176 p., spec. p. 81.

[69] More information about this label is available at https://beterleven.dierenbescherming.nl/english/.

[70] This label relies on 230 animal welfare indicators and ranks products on a five-level scale that goes from A (superior animal welfare) to E (minimal animal welfare standards). The first three levels, A, B, and C, recognize best practices that reflect significant improvement in the treatment of farm animals, while levels D and E guarantee compliance with minimum regulatory practices. Independent, third-party certifiers control labeled farms. See Mounier, L. (coord.). 2021. *Le bien-être des animaux d'élevage. Évaluer le bien-être animal.* Versailles: éditions Quae, coll. "Les mémos." See also Casino, Etiquetage: un pas de plus pour informer le consommateur sur le Bien-Être animal. Available at:https://www.groupe-casino.fr/etiquetage-un-pas-de-plus-pour-informer-le-consommateur-sur-le-bien-etre-animal/. Accessed 7 December 2024.

[71] Grimonprez, B. 2019. Le bien-être des animaux d'élevage: mythe ou réalité juridique? *Dictionnaire permanent Entreprise agricole*: 1–14.

[72] Di Concetto, A. and A.-S. Epstein. 2023. EU Consumer Information as a Tool to Regulate the Treatment of Farm Animals. *EU Journal of Consumer Law*: 145–174.

4 Conclusion

The agro-ecological transition, or rather revolution, is necessary and urgent, but also economically, socially, and politically costly, as well as difficult to define and implement. An effective policy would be to draw from the early years of the EU's Common Agricultural Policy, which provided strong agro-ecological measures.[73] The green agricultural revolution, rather than being prepared in ministerial cabinets, should take place in small steps, mobilizing all the resources of the law, the courts, and markets. Using consumer law to further animal protection goals could be one way of achieving such a revolution.

[73] On the CAP, also see Chap. 3, Nicolas de Sadeleer and Alice Di Concetto, "The Common Agricultural Policy and Animals: Friends or Foes?"

Meat Reduction as an Animal Protection Policy Goal: Exploring the Economic, Health, and Climate Nexus in EU Law

Marine Friant-Perrot

1 Introduction

Per capita meat consumption in France has now reached 84.9 kilograms annually, which is an alarming figure[1] not limited to France. Similarly high levels of meat consumption can be found throughout the whole EU, with

[1] Gérard, M. 2023. Un Français consomme deux fois plus de viande que la moyenne mondiale. *Le Monde*, 1 June. Available at: https://www.lemonde.fr/planete/article/2023/06/01/un-francais-consomme-deux-fois-plus-de-viande-que-la-moyenne-mondiale_6175784_3244.html. Accessed 22 December 2024.

M. Friant-Perrot (✉)
Law and Social Change Laboratory, UMR-CNRS 6297-Nantes University, Nantes, France
e-mail: marine.friant-perrot@univ-nantes.fr

A.-S. Epstein, A. Di Concetto (eds.), *Animals in EU Economic Law*, The Palgrave Macmillan Animal Ethics Series,
https://doi.org/10.1007/978-3-032-16171-0_11

the highest consumers of meat being the Spanish, followed by the Portuguese.[2]

Such high levels of meat consumption are deleterious for the planet. A 2013 report by the Food and Agriculture Organization (FAO) of the United Nations reported that food production is responsible for between 20% and 30% of all greenhouse gas emissions, with meat production alone accounting for 14.5% of emissions, more than the transport sector.[3] More specifically, one kilogram of meat produces five to ten times more greenhouse gases than a kilogram of cereals. A reduction in meat consumption and animal-based products is therefore crucial to achieve climate targets. For instance, in France, all carbon neutrality scenarios assume a reduction of animal-based food consumption ranging from at least 20%[4] up to 70%.[5] Worldwide, the EAT Foundation and *The Lancet* recommend reducing red meat consumption by an average of 50%, with an 84% reduction target for North Americans and a 77% reduction target for Europeans.[6]

While meat can be a reliable source of protein, massive meat consumption, particularly red and processed meat, is also detrimental to human health, as reported in a 2023 report by the World Health Organization (WHO).[7] More specifically, high consumption levels of red and processed meat increase the incidence of non-communicable diseases, including cancer, cardiovascular disease, and type 2 diabetes. For this reason, the European Commission has recognized that "current food consumption patterns are unsustainable from both health and environmental points of view" and that "in the EU, average intakes of [...] red meat [...] continue

[2] OECD. Indicators: Meat Consumption. Available at: https://www.oecd.org/fr/data/indicators/meat-consumption.html.

[3] Gerber, P.J., Steinfeld, H., Henderson, B., Mottet, A., Opio, C., Dijkman, J., Falcucci, A., and G. Tempio. 2013. Tackling climate change through livestock—A global assessment of emissions and mitigation opportunities. Food and Agriculture Organization of the United Nations (FAO), Rome. Available at: http://www.fao.org/3/a-i3437e.pdf.

[4] Stratégie nationale bas carbone (SNBC). Available at: https://www.ecologie.gouv.fr/politiques-publiques/strategie-nationale-bas-carbone-snbc.

[5] French Agency for Ecological Transition (ADEME). Available at: https://www.ademe.fr/les-futurs-en-transition/les-scenarios/.

[6] Willett, W., Rockström, J., Loken, B., Nestle, M., Lang, Y., Vermeulen, S., DeClerck, F. et al. 2019. Food in the Anthropocene: the EAT-Lancet Commission on healthy diets from sustainable food systems. *The Lancet* 393(10170): 447–492.

[7] WHO. 2023. *Red and processed meat in the context of health and the environment: many shades of red and green: information brief.* World Health Organization. ISBN: 978-92-4-007482-8.

to exceed recommendations" in its 2020 Farm to Fork Strategy[8] and 2021 European Beating Cancer Plan.[9]

Consumers are aware of the detrimental effects of meat consumption, even though only a small minority of them choose to avoid eating meat. Only 2.2% of the French population reported following a vegetarian diet in 2020,[10] and those who did claim to abstain from meat mainly for ethical reasons in support of animal protection, as well as health and environmental considerations.[11] However, close to 25% of French people claim to be flexitarians,[12] a dietary practice that consists of reducing meat consumption, primarily for health reasons. Regardless of the motives, these new dietary practices aid meat reduction targets and protein transition, in a way that seems to contradict the more general trend of meat consumption levels, which are now on the rise after several years of steady decline (+1% in 2021). This rise has been driven by an increased consumption of poultry,[13] processed products, and pre-packaged food products, in the context of an increasing number of meals eaten outside the home.[14]

The necessity of reducing meat consumption also implicates producers, since production patterns are influenced, at least partially, by consumption patterns. A challenge for producers is to reconcile two policy objectives that are at odds with one another: incentivizing the shift to more plant-based diets, while maintaining high production levels of meat. Two recent

[8] European Commission. 2020. Communication from the Commission to the European Parliament, the Council, the European Economic and Social Committee and the Committee of the Regions A Farm-to-Fork Strategy for a Fair, Healthy and Environmentally-Friendly Food System. COM (2020) 381 final, p. 13.

[9] European Commission. 2021. Communication from the Commission to the European Parliament and the Council, European plan to beat cancer. COM(2021) 44 final, p. 13.

[10] 2.2% of French people are vegetarians, according to a survey commissioned from Ifop by France AgriMer. Available at: https://www.franceagrimer.fr/content/download/66748/document/RapportCompletVegetariensFlexitariensEnFrance2020.pdf.

[11] Bertrandias, L., Cazes-Valette, G., and P. Gurviez. 2021. La préoccupation du bien-être animal, quels effets sur la consommation de viande? *Décisions Marketing* 2021/3 (103): 83–105 (in French).

[12] FranceAgriMer. 2021. Végétariens et flexitariens en France en 2020. Available at: https://www.franceagrimer.fr/Actualite/Etablissement/2021/VEGETARIENS-ET-FLEXITARIENS-EN-FRANCE-EN-2020 (in French).

[13] However, beef remains the second most consumed meat in France after pork. See Agreste. 2022. La consommation de viande en France en 2021, no. 394.

[14] FranceAgriMer. 2023. La consommation de viandes en France en 2022. Synthèses, n° 2023 412. Available at: https://agreste.agriculture.gouv.fr/agreste-web/download/publication/publie/SynCsm23412/consyn412202307-ConsoViande.pdf (in French).

examples of attempts to reduce meat production illustrate the tension between these objectives. In the Netherlands, the attempt to decrease national meat production volumes has led to a lasting political crisis, which originated in a 2019 ruling by the Dutch Council of State (the highest administrative court) in a case brought by environmental organizations against the Dutch government. Following a preliminary ruling by the Court of Justice of the EU,[15] the Dutch Council of State ruled that the government's nitrogen action plan was not ambitious enough to bring the Netherlands into alignment with conservation measures in Natura 2000 areas designated by the EU's Habitats Directive, and violated EU law as a result.[16] In response, the Dutch government published a "Nitrogen Action Program" in 2022, which aimed to reduce the number of farm animals by 30% to bring the Netherlands in compliance with EU law.[17] The publication of the program provoked farmers' anger and violent protests, with lasting effects on Dutch and EU politics. Between 2019 and 2024, the political party "Dutch Farmer-Citizen Movement" (BoerBurgerBeweging–BBB), which was created in the wake of the farmers' protest, won 23 seats in the Dutch parliament and two seats in the European Parliament.

Similarly in France, a 2023 report by the French Court of Auditors recommended reducing the number of cows to reduce CO2 emissions, and more specifically, the Court proposed that the government should "define and make public a strategy for reducing" the number of cows farmed in France.[18] The auditors primarily sought to reduce the environmental footprint of animal agriculture, and since beef farming is a major source of greenhouse gases, the report did not address other sectors in animal agriculture, such as hog and poultry farming, which emit fewer greenhouse gases even though they pose more animal welfare risks com-

[15] Joined Cases C-293/17 and C-294/17, *Coöperatie Mobilisation for the Environment UA, Vereniging Leefmilieu v College van gedeputeerde staten van Limburg, College van gedeputeerde staten van Gelderland* (C-293/17), *Stichting Werkgroep Behoud de Peel v College van gedeputeerde staten van Noord-Brabant* (C-294/17), 2018 ECLI:EU:C:2018:882.

[16] Dutch Council of State. 2019. Decision 201600614/3/R2, 201600617/3/R2, 201600618/3/R2, 201600620/3/R2, 201600622/4/R2, 201600630/3/R2; ECLI:NL:RVS:2019:1603. Available at: https://www.raadvanstate.nl/uitspraken/@115602/201600614-3-r2/.

[17] United States Department of Agriculture, Foreign Agricultural Services. 2021. Dutch Parliament Approves Law to Reduce Nitrogen Emissions; Vallet, C. 2022. Aux Pays-Bas, un plan drastique de réduction de l'azote provoque la colère des agriculteurs. *Le Monde*, 13 July.

[18] Cour des Comptes. 2023. Les soutiens publics aux éleveurs de bovins. Available at: https://www.ccomptes.fr/fr/publications/les-soutiens-publics-aux-eleveurs-de-bovins.

pared to extensive cattle farming. However, because France is the top beef producer and the second largest dairy producer in the EU, the French government was prompt to reassure beef producers following the publication of the Court of Auditors' report. Prime Minister Élisabeth Borne asserted before the lower house (Assemblée Nationale) that livestock farming "has a future" in France, despite climate change, and that "livestock farming can have beneficial effects on the environment." Prime Minister Borne further stated that "every year, grasslands store 8 million tonnes of CO_2" and that livestock farming "enables the construction of a network of hedges and agro-ecological infrastructures that are particularly useful for the soil," whereas the Court of Auditors' report pointed out that carbon sequestration by grazing meadows was "far from offsetting [cattle] emissions."

These two examples show how challenging it is for governments in Europe to adopt and implement meat reduction policies in the face of powerful vested interests. As a result, economic regulations in the agri--food sector have incorporated non-economic objectives only to a marginal extent. The strong support for animal agriculture and the absence of any regulation in meat consumption volumes has thus paved the way for a regulatory framework that undermines meat reduction efforts. However, EU law offers promising avenues to change the paradigm in animal-source food production and consumption systems.

2 How Economic Regulations in the Agri-Food Sector Undermine Meat Reduction Objectives

Consistent with a pro-liberal trade agenda, EU law does not include any measure that aims to reduce meat production and consumption. Instead, the EU institutions have only implemented non-binding instruments, such as incentives and educational tools, that fail to question the productivist nature of animal agricultural production. Such instruments also deprioritize environmental, animal, and public health protection goals. As a result, the laws governing meat production and consumption primarily support the production of massive volumes of meat, and only take into account environmental and public health considerations to a minimal extent.

2.1 *Legal and Regulatory Instruments in Support of High Consumption Volumes of Meat*

In the EU, there are no meat reduction policy actions addressing the supply side or the demand side of animal agriculture.

2.1.1 *Common Agricultural Policy Instruments*

Historically, since World War II, the Common Agricultural Policy (CAP) regulations, which establish the rules on the distribution of agricultural subsidies, have aimed to produce large volumes of food to guarantee food security and self-sufficiency in Europe.[19] As per the EU's constitutional treaty, the Treaty on the Functioning of the EU (TFEU), "[increasing] agricultural productivity"[20] has been the primary objective of the CAP, in a way that has stimulated the production of animal-source food over the past six decades. Consistent with the logic that animals are only approached as "a factor of production and commodities,"[21] the TFEU classifies animals as "agricultural products."[22] The adoption of Article 13 TFEU, which recognizes animals as "sentient beings" and calls on the Union and the Member States "to pay full regard to the welfare requirements of animals," and EU animal welfare statutes, has not substantially altered this view rooted in the commodification of animals.[23] Such a situation can be

[19] Blumann, C., ed. 2011. *Politique agricole commune et politique commune de la pêche*. Brussels: Éditions de l'Université de Bruxelles; Bianchi, D. 2012. *La politique agricole commune—Précis de droit agricole européen*, p. 51. Bruxelles: Bruylant. For more on the Common Agricultural Policy, also see Chap. 3, Nicolas de Sadeleer and Alice Di Concetto, "The Common Agricultural Policy and Animals: Friends or Foes?"

[20] TFEU, art 39.

[21] Del Cont, C. 2021. L'animal d'élevage dans le droit de l'Union européenne ou la difficile conciliation de l'objectif de bien-être animal et des objectifs économiques. *Revue de l'Union européenne*, p 546; Sowery, K. 2018. Sentient Beings and Tradable Products: the Curious Constitutional Status of Animals Under Union Law. *Common Market Law Review* 17; Di Concetto, A. 2022. The Double Edged Sword: International Law and Its Effects on EU Farm Animal Welfare Legislation. *Global Journal of Animal Law, Special Issue: Internal Law and Animal Health and Protection: Persistent Themes, New Prospects for Change* 10(2); Di Concetto, A. 2024. EU Animal Law. In *Nordic Animal Law: Welfare and Rights*, ed. Birgitta Wahlberg. Cambridge: Ethics International Press.

[22] TFEU, art 38(1) and Annex I.

[23] Di Concetto, A. 2022. The Double Edged Sword: International Law and Its Effects on EU Farm Animal Welfare Legislation. *Global Journal of Animal Law, Special Issue: Internal Law and Animal Health and Protection: Persistent Themes, New Prospects for Change* 10(2); Di Concetto, A. 2024. EU Animal Law. In *Nordic Animal Law: Welfare and Rights*, ed. Birgitta Wahlberg. Cambridge: Ethics International Press.

explained by the fact that "animal welfare" as a concept was primarily conceived with a view to exploit animals and institutionalize suffering.[24] French legal scholar Sonia Desmoulin further underlines that the concept of animal welfare was "invented and subsequently used as concept with regulatory potential, with the aim of measuring animals' ability to adapt to a constrained and even oppressive environment to meet economic requirements" and that "[t]he animal whose welfare is assessed is an agricultural product resulting from a system that attempts to extract most economic value out of it."[25]

The EU legislature's neglect of farmed animals' interests, along with antiquated interpretation of the CAP objectives,[26] has persisted until today, as illustrated in the policy orientations set forth in the 2021 CAP regulations. Firstly, CAP objectives and payments fail to include a goal of reducing meat and animal production, as underscored by the European Court of Auditors in a 2021 report.[27] Another 2024 study further revealed that 82% of CAP subsidies end up supporting animal-source food production in the EU.[28] Besides the subsidies from which livestock farmers benefit directly ("coupled payments"), an even greater proportion of subsidies are allocated to animal feed production ("direct payments"), which likely has the effect of lowering production costs in the animal agriculture sectors.[29] A disproportionate amount of the CAP's budget also supports beef production, despite the sector's deleterious impact on the climate and

[24] Fabien, M. 2018. La protection du bien-être de l'animal par l'Union européenne. *Revue Trimestrielle de Droit européen* 2018/2, p. 251. For more on the concept of animal welfare, also see Chap. 10, Isabelle Doussan, "Using Consumer Law to Improve the Treatment of Farm Animals in Industrial Farming Production."

[25] We translated Desmoulin, S. 2021. Des difficultés révélatrices d'une véritable aporie? *Revue de l'Union européenne*, p. 567.

[26] See Bluman, C. 2023. Ecologisation de la PAC: nouvelle illustration de la politique des petits pas. *Revue de droit rural*: 26–32.

[27] European Court of Auditors. 2021. Special Report 16/2021, Common Agricultural Policy and Climate, p 20.

[28] Kortleve, A.J., Mogollón, J.M., Harwatt, H., and P. Behrens. 2024. Over 80% of the European Union's Common Agricultural Policy supports emissions-intensive animal products. *Nat Food* 5: 288–292.

[29] Le bien-être animal dans la nouvelle Politique Agricole Commune. 2023. *Revue de droit rural* n°1/2023; Kortleve, A.J., Mogollón, J.M., Harwatt, H., and P. Behrens. 2024. Over 80% of the European Union's Common Agricultural Policy supports emissions-intensive animal products. *Nat Food* 5: 288–292. The article states that, among the €46 billion spent on animal production, €21 billion are allocated to livestock feed and €18 billion to production itself.

public health. For instance, in France, the beef sector receives close to €4.3 billion every year, making it the most heavily subsidized agricultural sector, even though beef farms account for as little as one third of all farms and take up one third of France's arable lands.[30]

Secondly, CAP payments only include animal welfare measures to a minimal extent.[31] Although CAP regulations penalize producers by reducing their payments in cases of non-compliance with animal welfare laws,[32] such an enforcement mechanism has no effect on limiting intensive production practices in EU animal agriculture. This is because payments subject to penalties are based on the number of hectares producers have, and many producers in animal agriculture do not own many hectares (unlike crop growers). Other CAP payments in support of animal welfare, the so-called Rural Development Measures and eco-schemes, are allocated to producers who voluntarily undertake good environmental and animal welfare practices, which is not enough to effectively reduce production levels across animal agriculture.[33] As an alternative to these half-measures, legal scholars have thus recommended rendering producers who cannot demonstrate climate and environmental neutrality ineligible for subsidies.[34] This ineligibility measure could be extended to producers who do not comply with high animal welfare standards.

These rules in the distribution of subsidies are thus inconsistent with a "less but better" approach, which is needed in light of climate and sustainability priorities. While animal agriculture relies, to a certain extent, on

[30] Cour des Comptes. 2023. Les soutiens publics aux éleveurs de bovins. Available at: https://www.ccomptes.fr/fr/publications/les-soutiens-publics-aux-eleveurs-de-bovins.

[31] Di Concetto, A. 2023. Le bien-être animal dans la Politique Agricole Commune: la prise en compte d'une "attente sociétale." *Revue de droit rural* N°1: 14–19. For a more detailed analysis, see Chap. 3, Nicolas de Sadeleer and Alice Di Concetto, "The Common Agricultural Policy and Animals: Friends or Foes?"

[32] Regulation (EU) n°2021/2115. For a presentation of this text, see Bluman, C. 2023. Ecologisation de la PAC: nouvelle illustration de la politique des petits pas. *Revue de droit rural*: 26–32.

[33] Di Concetto, A. 2023. Le bien-être animal dans la Politique Agricole Commune: la prise en compte d'une "attente sociétale." *Revue de droit rural* N°1: 14–19. For a more detailed analysis, see Chap. 3, Nicolas de Sadeleer and Alice Di Concetto, "The Common Agricultural Policy and Animals: Friends or Foes?"

[34] Bodiguel, L. 2020. Building a new common sustainable agricultural and food legal model in the face of the climate and food emergency: from transition to mutation. *Revue européenne de droit de la consommation* 2020/1: 29–42.

subsidies, the transition to sustainable animal agricultural systems requires a paradigm shift in the allocation and distribution of subsidies.

2.1.2 Food Information to Consumer Legislation

On the consumption side of the agri-food value chain, EU legislation provides no rules that restrict levels of animal-source food consumption. A general principle of EU law and policy is that each individual is free to choose their diet as they see fit. Consumers' freedom of choice is positive and banning meat consumption would likely be at odds with the fundamental rights of consumers and producers. However, this freedom of choosing one's eating habits is in fact limited by a number of factors, starting with advertising campaigns in support of meat-based diets. Furthermore, public health objectives provide solid grounds for regulating meat consumption levels, especially red meat, given its central role in causing disease.

Under current EU legislation, health risks caused by excessive consumption of fatty and sugary foods are regulated through the implementation of preventive measures.[35] A central measure in that respect has been the introduction of mandatory nutrition labeling, required under Regulation 1169/2011 on Food Information to Consumers (FIC Regulation).[36] The European Commission—the EU's executive branch—

[35] On the EU's competence in regulating public health, see Garde, A. 2010. EU Law and Obesity Prevention. *Kluwer Law International*; Bartlett, O., and A. Garde. 2017. The EU's Failure to Support Member States in their Implementation of the WHO Recommendations: How to Ignore the Elephant in the Room? *European Journal of Risk Regulation* 8(2): 251–269. https://doi.org/10.1017/err.2017.30; Delhomme, V.N. 2020. Emancipating health from the internal market: for a stronger EU (Legislative) competence in public health. *European Journal of Risk Regulation* 11(4—Special issue "Beyond COVID-19: Towards a European Health Union"): 747–756. On the specific topic of the EU's competence in regulating lifestyle-related health risks, including dietary risks, see Friant-Perrot, M., and A. Garde. 2023. Le défi d'une alimentation plus saine: l'Union et la lutte contre l'obésité et le surpoids. In *Les 20 ans du règlement sur la législation et la sécurité alimentaire dans l'Union européenne—Bilan et perspectives*, Nathalie De Grove-Valdeyron and Marine Friant-Perrot, eds. Bruxelles: Bruylant; Delhomme, V.N. 2024. Minimum harmonization, experimentation and the internal market: balancing unity and diversity in the regulation of lifestyle-related health risks. In *Balancing Unity and Diversity in EU Legislation*, Ton van den Brink and Virginia Passalacqua, eds. 194–210. Cheltenham: Edward Elgar.

[36] Regulation n°1169/2011 on the provision of food information to consumers, amending Regulations (EC) No 1924/2006 and (EC) No 1925/2006 of the European Parliament and of the Council, and repealing Commission Directive 87/250/EEC, Council Directive 90/496/EEC, Commission Directive 1999/10/EC, Directive 2000/13/EC of the

had also announced the introduction of a sustainable food label in 2020 as a way to raise consumer awareness about sustainable food consumption.[37] This food label could have been a vehicle to include animal welfare standards, but the Commission has since abandoned this proposal.[38]

Overall, however, the transparency principles that are central in EU food law still fail to address high levels of meat consumption, and the deleterious effect such levels have on animals.[39] For instance, EU law requires manufacturers to display the presence of animal-source food in the denomination of raw products, as well as the presence of animal-source food in the ingredient list of processed food products. However, these rules do not require manufacturers to specify the proportion of such animal-source products in raw and processed products.[40] Such rules therefore may skew flexitarians' purchasing choices. Yet, the EU Legislature has the authority to introduce specific consumer information to inform consumers of recommended consumption levels of animal-source food products, or to recommend reducing such consumption, or even disincentivizing such consumption.

European Parliament and of the Council, Commission Directives 2002/67/EC and 2008/5/EC and Commission Regulation (EC) No 608/2004, 2011 O.J. (L 304) 18–63.

[37] Measure 23 of Communication from the Commission to the European Parliament, the Council, the European Economic and Social Committee and the Committee of the Regions A Farm to Table Strategy for a Fair, Healthy and Environmentally Friendly Food System; Di Lauro, A. 2023. La stratégie de la ferme à la table: vers plus de transparence sur les impacts écologiques des denrées alimentaires? In *Les 20 ans du règlement sur la législation et la sécurité alimentaire dans l'Union européenne—Bilan et perspectives*, Nathalie De Grove-Valdeyron and Marine Friant-Perrot, eds. Bruxelles: Bruylant. Besides health considerations, consumer law can be used as a vehicle to achieve environmental objectives. For more on this, see Sauphanor-Brouillaud, N. 2022. Le consommateur, levier de la transition écologique. In *Le droit économique, levier de la transition écologique*, Aude-Solveig Epstein and Marie Nioche, eds. 137–154. Bruxelles: Bruylant.

[38] The European Commission mentioned the idea of an animal welfare label again in 2025 in the context of its Vision on Agriculture and Food, though using unclear terms. Source: European Commission. 2025. Communication from the Commission to the European Parliament, the Council, the European Economic And Social Committee and the Committee of the Regions: A Vision for Agriculture and Food Shaping Together an Attractive Farming and Agri-Food Sector for Future Generations. COM(2025) 75 final, p.24.

[39] Bouillot, P.E. 2020. Enjeux juridiques de la transition protéique: le droit de la consommation aveugle aux enjeux de l'alimentation durable? *R.E.D.C.* 2020/1: 87–100.

[40] Articles 9 and 18 Regulation1169/2011 on the provision of food information to consumers, O.J. (L 304) 18–63. Unless the sales name mentions this ingredient or the presentation specifically highlights it. For example, beef lasagne with an indication of the percentage of beef (e.g. 15%) (art. 22).

Existing consumer information rules do not adequately address high meat consumption levels. The mandatory country-of-origin labeling on all meat products, which has been a requirement since the BSE ("mad cow") crisis for beef products in 2000 and extended to all meat products in 2013,[41] was primarily motivated by the necessity to restore consumers' confidence in EU meat production and stimulate consumption, as opposed to increasing transparency and incentivizing sustainable food choices. As a result, country-of-origin labeling is not associated with information on farming methods, and so is not adequate to guide consumers from a sustainability or animal welfare perspective.[42]

Additionally, EU law provides no rules regarding the labeling of meat-free food products, such as vegetarian and plant-based products. The FIC Regulation provided that the European Commission should adopt a regulatory definition for vegan and vegetarian, by specifying "information related to suitability of a food for vegetarians or vegans."[43] However, the Commission never adopted such an implementing act. A 2024 report by the European Court of Auditors therefore identified the lack of EU rules harmonizing vegetarian and vegan labels as a "notable gap," highlighting that "in the absence of EU rules for such food products, consumers can only base their decisions on the different private labels and product names."[44]

In France, the issue of food information to consumers on vegetarian and plant-based products has given rise to a contentious debate around food and identity politics. In this context, the French government has attempted to secure the political support of animal agriculture groups, while also promoting more plant-based eating habits. Such contradictory

[41] Regulation (EC) No 1760/2000 of the European Parliament and of the Council of 17 July 2000 establishing a system for the identification and registration of bovine animals and regarding the labelling of beef and beef products and repealing Council Regulation (EC) No 820/97, 2000 O.J. (L 204) 1–10; Commission Implementing Regulation (EU) No 1337/2013 of 13 December 2013 laying down rules for the application of Regulation (EU) No 1169/2011 of the European Parliament and of the Council as regards the indication of the country of origin or place of provenance for fresh, chilled and frozen meat of swine, sheep, goats and poultry, 2013 O.J. (L 335) 19–22.

[42] For more on food labeling, see Chap. 10, Isabelle Doussan, "Using Consumer Law to Improve the Treatment of Farm Animals in Industrial Farming Production."

[43] Regulation 1169/2011 on the Provision of Food Information to Consumers, art 36(3)(b), 2011 O.J. (L 304) 37.

[44] European Court of Auditors. 2023. Special Report 23/2024: Food Labelling in the EU—Consumers Can Get Lost in the Maze of Labels, p. 26 and 43.

political messages have resulted in the adoption of controversial laws. For instance, in 2022, the French administration adopted a rule regulating the use of terms for plant-based food products,[45] which prohibited manufacturers from using names traditionally used for animal-based products, such as "steak" or "sausages."[46] The plant-based protein trade association *Association Protéines France* challenged the legality of the rule before France's highest administrative court (Conseil d'Etat). In a 2022 emergency decision, which did not rule on the merits, the Court found the rule to be unlawful on the grounds that it infringed upon the principle of legal certainty.[47] In a subsequent ruling on the merit, handed down in 2023, the French Court referred a series of questions to the Court of Justice of the European Union (CJEU) to obtain a preliminary ruling, aiming to clarify the legal status of these designations under EU law.[48]

However, the French administration did not wait for the Court's ruling and prepared a new draft rule reserving all traditional terms for animal-based products, thereby prohibiting the use of such terms for plant-based alternatives. In 2023, the French administration notified the European Commission of the draft rule[49] and subsequently adopted the rule in February 2024,[50] but the Conseil d'Etat suspended its implementation in April 2024.[51] The CJEU eventually ended this judicial saga by a ruling handed down in October 2024, in which the Court ruled against France and determined that Member States cannot prohibit the use of terms traditionally linked to animal-based products for the labeling of products

[45] This decree specifies the terms of application of article L. 412–10 inserted into the consumer code by article 5 of the law of 10 June 2020 on the transparency of information on agricultural and food products, which states that "The names used to designate foodstuffs of animal origin may not be used to describe, market or promote foodstuffs containing plant proteins."

[46] Bombardier, J. 2022. Denrées alimentaires d'origine végétale: quel encadrement pour les termes "lait", "steak" et "fromage" utilisés dans la dénomination ou la présentation d'un produit végétal. *Revue de droit rural* n°505.

[47] Conseil d'État, 27 July 2022, n°465844 (in French).

[48] Conseil d'État, 12 July 2023, n°465835 (in French).

[49] France. 2023. Notification to the Commission, Décret relatif à l'utilisation de certaines dénominations employées pour désigner des denrées comportant des protéines végétales, 2023/0510/FR.

[50] Decree no. 2024–144 of 26 February 2024 on the use of certain names used to designate foodstuffs containing plant proteins, JORF no. 0048 of 27 February 2024.

[51] Conseil d'État, 10 April 2024, n°492,844 (in French).

containing vegetable proteins—unless specifically restricted by EU law.[52] The 2024 CJEU ruling is consistent with its previous case law. In a 2017 ruling,[53] the CJEU had determined that the terms "milk" and "dairy product" could not be used for a plant product because EU law specifically reserved these terms for animal-based products, as per Regulation 1308/2013 on the Common Organization of the Markets (CMOs Regulation).[54] In its 2024 ruling, the CJEU thus considered that there was no such EU harmonized legislation establishing reserved terms for plant-based proteins: "EU law does not lay down any rule reserving for certain foods, specifically defined as being of animal origin, the use of legal names containing terms from the butchery, charcuterie and fish sectors covered by Decree No 2022-947 or by the names listed in Annex I to Decree No 2024-144."[55]

EU law is in fact unambiguous on the use of terms for plant-based products. In the absence of specific rules (as is the case with milk and dairy products), the FIC Regulation provides that the "descriptive name" of a food product is "a name providing a description of the food, and if necessary of its use, which is sufficiently clear to enable consumers to know its true nature and distinguish it from other products with which it might be confused."[56] The CJEU in 2024 specifically ruled that Member States were not allowed to regulate these denominations on account of fast-track procedure harmonizing denomination rules, in light of the definitions of these denominations, as well as articles 7 (on the prohibition to mislead consumers) and 9 (on mandatory information), and Annex VI (A)(4) (on the mandatory particulars accompanying the name of the food in the event of a substitution of an ingredient or component). In this context, said the Court, consumers should not be deprived of information related

[52] C-438/23, *Protéines France, Union végétarienne européenne (EVU), Association végétarienne de France (AVF), Beyond Meat Inc. v Ministre de l'Économie, des Finances et de la Souveraineté industrielle et numérique*, 4 October 2024.

[53] CJEU, C-422/16, *Verband Sozialer Wettbewerb eV v. TofuTown.com GmbH*, 2017 ECLI:EU:C:2017:458; See also, in this sense, CJCE, C-101/98, *UDL*, 1999 EU:C:1999:615.

[54] Regulation 1308/2013 establishing a common organisation of the markets in agricultural products, art 78(1)(c) and Annex VII, Part III, 2013 O.J. (L 347) 715 and 814–815.

[55] C-438/23, *Protéines France, Union végétarienne européenne (EVU), Association végétarienne de France (AVF), Beyond Meat Inc. v Ministre de l'Économie, des Finances et de la Souveraineté industrielle et numérique*, 4 October 2024, para. 78.

[56] Regulation (EU) No 1169/2011 of the European Parliament and of the Council of October 25, 2011 on the provision of food information to consumers, art 2(p), O.J. (L304) 25.

to common use or descriptions. Consistent with this standard, the Rennes Court of Appeal had ruled in 2022 that the names "steak" and "sausages" did not necessarily mislead consumers when used to designate plant-based food products.[57] The European Commission's written observations to France following the notification of the French rule in 2023 also supported this view, as the Commission considered that food manufacturers had used some of the terms banned in the French rule on the EU market for several years, and that consumers had become familiar with these terms.[58]

EU law thus reflects cultural eating habits, in which meat consumption is a cultural marker, with limited space for protein transition.

2.2 The Absence of Robust Meat Reduction Law and Policy Instruments to Address Public Health and Environmental Issues

Like EU agricultural and consumer information legislation, EU health and environmental legislation falls short of limiting high meat consumption levels. Few Member States have attempted to fill this shortcoming with the adoption of national policies. Key national policy measures are the official nutritional guidelines and the generalization of meat-free options in school cafeterias, even though these measures carry little weight in the face of promotional campaigns in favor of meat consumption.

2.2.1 National Official Nutritional Guidelines

Likely on account of the EU's limited competence in public health, there are no EU nutritional guidelines. However, virtually all Member States have adopted national nutritional guidelines. These guidelines are not codified in law and typically function as non-binding policy initiatives aimed at educating and incentivizing the general public to adopt a healthy diet.

In France, nutritional guidelines have regularly evolved since 2001. The most recent version of the French nutritional guidelines was adopted

[57] Bombardier, J. 2022. Denrées alimentaires d'origine végétale: quel encadrement pour les termes "lait", "steak" et "fromage" utilisés dans la dénomination ou la présentation d'un produit végétal. *Revue de droit rural* n°505.

[58] European Commission. 2023. Communication from the Commission—Decree on the use of certain names to designate foodstuffs containing vegetable proteins, TRIS/(2024) 0886, 22 November.

in 2019,[59] and these recommend reducing meat consumption for both health and environmental reasons.[60] More specifically, French nutritional guidelines recommend that the general population eat poultry while also limiting intakes of all types of meat (including poultry) to 500 grams per week per person.[61] The guidelines further recommend limiting consumption of dairy products to two portions per day for adults and three for children.[62] Similar recommendations exist in other European countries. For example, Finnish authorities advise limiting meat consumption to less than 500 grams per week.[63] Other European countries recommend even lower meat level consumption, such as Norway[64] and Denmark,[65] where the suggested weekly maximum is 350 grams.

These nutritional guidelines are disseminated to the general population in a variety of ways, such as nutrition guides, a dedicated public website,[66] and media campaigns. These guidelines are also disseminated by health care, social, and education professionals. French law further requires advertisers and manufacturers to include health messages based on official nutritional guidelines.[67] However, even when meat industry groups integrate some of these recommendations, by promoting flexitarianism, for

[59] Ministère de la Santé. Programme National Nutrition Santé 2019–2023, p. 39–41. Available at: https://sante.gouv.fr/IMG/pdf/pnns4_2019-2023.pdf. The 4th PNNS has been extended to 2024 to better align with the future National Strategy for Food, Nutrition, and Climate (SNANC) for 2030 and the next PNNS for 2025–2030.

[60] Other official nutritional guidelines in the Member States include recommendations on more plant-rich diets, such as the Finnish guidelines.

[61] Santé publique France. 2019. Recommandations sur l'alimentation, l'activité physique et la sédentarité pour les adultes. Available at: https://www.santepubliquefrance.fr/presse/2019/sante-publique-france-presente-les-nouvelles-recommandations-sur-l-alimentation-l-activite-physique-et-la-sedentarite.

[62] Ibid.

[63] Finnish Food Authority. Nutrition recommendations and food-based dietary guidelines. Available at: https://www.ruokavirasto.fi/en/foodstuffs/healthy-diet/food-groups/meat-meat-products-and-eggs/. Accessed 7 April 2025.

[64] The Norwegian Directorate of Health. Norway's nutritional guidelines. Available at: https://www.helsenorge.no/en/kosthold-og-ernaring/dietary-guidelines/. Accessed 7 April 2025.

[65] Ministry of food, agriculture and fisheries of Denmark. The Danish Official Dietary Guidelines. Available at: https://en.fvm.dk/news-and-contact/focus-on/the-danish-official-dietary-guidelines. Accessed 7 April 2025.

[66] Santé publique France. Recettes, idées d'activités et astuces pour manger mieux et bouger plus petit à petit. Available at: https://www.mangerbouger.fr/.

[67] Article L.2133-1 of the French Public Health Code.

instance, the messaging they use generally attempts to maintain high consumption levels by promoting "better" meat, as opposed to consuming less.[68]

Overall, official nutritional guidelines are not effective enough to counteract pro-meat speech[69] due to several limitations. A first limitation has to do with inconsistencies in the public budget allocated to public health and agricultural policies that aim to influence citizens' eating habits. For instance, the amount of the French national budget that is dedicated to preventative measures in public health represents as little as 2% of healthcare expenditures in France.[70] On the other hand, between 2016 and 2020, a third of the EU's budget for the promotion of agricultural products in the EU and abroad (776.7 million euros) was allocated to the promotion of meat and dairy products (252.4 million euros).[71] In 2020, the European Commission announced it would end such inconsistencies between public health and agricultural goals by reviewing the EU agri-food promotion program so as "to support the most sustainable, carbon-efficient methods of livestock production" in the Farm to Fork Strategy.[72] However, the selection rules for EU agri-food promotion programs have

[68] Bienvenu, J. 2019. Le "flexitarisme", un régime encore flou récupéré par le lobby de la viande. *Le Monde*, 1 March; Greenpeace. 2021. Que. cache la pub pour la viande? Available at: https://cdn.greenpeace.fr/site/uploads/2021/11/Rapport_Hache_Menu_Ce_que_cache_la_publicite_pour_la_viande_2.pdf?_ga=2.74362139.868852813.1692972829-1488030645.1692972829.

[69] Friant-Perrot, M. 2017. Enjeux juridiques de la promotion d'une alimentation saine: messages sanitaires et marketing alimentaire en France. In *Agir sur les comportements nutritionnels: Réglementation, marketing et influence des communications de santé*, ed. Inserm, 45–62. Montrouge: EDP Sciences.

[70] €7.5 billion out of €325 billion in 2023. Source: Didier, M. and G. Lefebvre, eds. 2024. Les dépenses de santé en 2023—Résultats des comptes de la santé—Édition 2024. Available at: https://drees.solidarites-sante.gouv.fr/publications-communique-de-presse-documents-de-reference/panoramas-de-la-drees/241120-Panorama-CNS24 (in French).

[71] Di Concetto, A. 2022. Les campagnes de promotion de viande et de lait: un changement notable d'orientation des politiques publiques européennes. La Fondation Droit Animal Ethique & Science, 28 February. Available at: https://www.fondation-droit-animal.org/112-campagnes-promotion-viande-lait-changement-notable-dorientation-politiques-publiques-europeennes/.

[72] European Commission. 2020. Communication from the Commission to the European Parliament, the Council, the European Economic and Social Committee and the Committee of the Regions A Farm-to-Fork Strategy for a Fair, Healthy and Environmentally-Friendly Food System. COM (2020) 381 final, p. 7–8.

remained the same between 2020[73] and 2025.[74] Over the period 2020–2024, 116 of the 220 promotional campaigns financed concerned animal products, at a cost of more than €230 million.[75] This represents more than 52% of the total budget for promotion programs, of which 33% is for meat, dairy products, and eggs and 19% is for product baskets containing products of animal origin. The distribution of funds under the agricultural products promotion program appears to be inconsistent with the Farm to Fork strategy's objective of transitioning toward more plant-based diets.

A second limitation of official nutritional guidelines is that private interest groups in the agri-food sector disproportionately benefit from favorable freedom-of-speech rules compared to public interest groups advocating for meat reduction. For example, the Swiss government denied the animal protection association "Verein Gegen Tierfabrieken" (Association Against Animal Factories) the authorization to broadcast a TV commercial encouraging consumers to "eat less meat" for "health, animal protection, and environmental protection" reasons. This commercial was developed in response to a series of pro-meat commercials by the meat industry; however, the Swiss authorities considered that the Verein Gegen Tierfabrieken's video clip was "manifestly political" and overly provocative. Verein Gegen Tierfabrieken subsequently sued the Swiss authorities, and after an eight-year legal saga, the European Court of Human Rights ruled in 2001 that the Swiss authorities had violated Article 10 on freedom of expression of the European Convention of Human Rights.[76]

[73] Commission implementing decision on the financing of information provision and promotion measures concerning agricultural products implemented in the internal market and in third countries and the adoption of the work programme for 2020, Annex I. COM/2019/8095 final. Available at: https://agriculture.ec.europa.eu/document/download/3ffb2351-8b13-419f-9b35-1c3b198d6f06_en?filename=annex-commission-decision-c2019-8095_en.pdf.

[74] Commission implementing decision on the financing of information provision and promotion measures concerning agricultural products implemented in the internal market and in third countries and the adoption of the work programme for 2025, Annex I. COM/2024/8679 final. Available at: https://webgate.ec.europa.eu/circabc-ewpp/d/d/workspace/SpacesStore/cd218792-1820-4ecb-87bc-0519de78e672/file.bin.

[75] Lists of selected programs available on: European Commission, European Research Executive Agency. Calls for Proposals—Promotion of Agricultural Products. Available at: https://rea.ec.europa.eu/funding-and-grants/promotion-agricultural-products-0/calls-proposals-promotion-agricultural-products_en. Accessed 28 April 2025.

[76] Cour EDH, *Verein Gegen Tierfabriken v. Suisse*, 2001 req. n° 24,699/94; Marguénaud, J.-P. 2017. Le droit européen des droits de l'Homme et la protection des animaux. *Revue des Affaires Européennes* 2017/1: 77–90.

As a result of budgetary inconsistencies and rules favorable to pro-meat speech, official nutritional guidelines have been largely ineffective in shifting eating habits toward more plant-based diets. In fact, the French administration generally does not assess the effectiveness of such guidelines. The mid-term evaluation of the 2019–2023 nutritional guidelines only included data on recommendations for vegetable consumption but did not assess the effect of recommendations to lower meat consumption in the general population.[77] However, official data by the French Ministry of Agriculture reports that total meat consumption in France significantly increased between 1970 and 2022.[78] The trend appeared to be shifting in 2023, as total per capita meat consumption had declined in France.[79] However, this decrease cannot be directly attributed to official nutritional guidelines and public policies aimed at reducing meat consumption levels. In fact, the trend was occurring in an inflationary context that was impacting consumers' purchasing power and conceals disparities in consumption patterns depending on the type of meat. While red meat consumption decreased, poultry meat consumption increased.[80]

2.2.2 *Vegetarian Alternatives in School Cafeterias as a Way to Promote Plant-Rich Diets*

In France, the legislature has been reluctant to implement vegetarian or plant-based options in school cafeterias, on account of cultural traditions.[81] Courts have also ruled that the French government does not have an obligation to provide vegetarian or plant-based options to children in schools, because the children are not constrained to the school cafeterias for their meals, unlike prisoners.[82]

[77] Ministère de la Santé. 2021. Mid-term review of the 2019–2023 National Nutrition and Health Program. Available at: https://sante.gouv.fr/IMG/pdf/pnns_4_bilan_mai2021.pdf.

[78] It rose from 3.8 million to 5.8 million tons of carcass equivalent between 1970 and 2022, a 50% increase. Source: FranceAgriMer. 2023. La consommation de viandes en France en 2022. Synthèses, n° 2023 412. Available at: https://www.franceagrimer.fr/content/download/71709/document/SYN-VIA-Conso%20viande%20Fce2022.pdf.

[79] FranceAgriMer. 2023. La consommation de viandes selon les profils socioprofessionnels et les régimes alimentaires. Synthèses, n° 424. Available at: https://www.franceagrimer.fr/content/download/74051/document/SYNTHESE_424_ConsoViande%20SSP-FAM.pdf.

[80] Ibid.

[81] Poulain, J.-P. 2012. *Modèle alimentaire. Dictionnaire des cultures alimentaires.* Paris: PUF, p. 881–888.

[82] Conseil d'État, Association végétarienne de France, 20 March 2013, n°354547 (in French).

However, the law in France has evolved under the influence of climate objectives, and the fact that almost half of the meals in France are eaten outside the home. By way of a 2018 law,[83] France enacted the obligation for all French school cafeterias to provide at least one vegetarian option per week. This law was first implemented on an experimental basis for two years (2019–2021). To encourage the transition to more plant-rich diets, the law further requires public catering services to promote plant-based alternatives to animal-based products through the adoption of a multi-year plan for protein diversification.[84] A 2021 law subsequently mandated all public and private cafeteria vendors to offer a vegetarian option on a daily basis.[85] However, a 2021 evaluation reported that 37% of local administrations were not complying with this law.[86] A 2023 evaluation showed progress in the implementation of the law, despite persisting challenges.[87]

While the generalization of vegetarian options in cafeterias has been a relative success, the adoption of additional and more robust transition measures in France is limited by a rule on the nutritional quality of meals served in school cafeterias.[88] This rule was adopted more than 20 years ago and contradicts official nutritional guidelines and best available nutritional science[89] by requiring school cafeterias to offer "proteins composed of

[83] Law no. 2018–938 of October 30, 2018 for balanced trade relations in the agricultural and food sector and healthy, sustainable food accessible to all, JORF no. 0253 of 1 November 2018, codified in Article L.230–5-6 of the French Rural Code.

[84] Conseil national de la restauration collective. 2021. Plan pluriannuel de diversification des sources de protéines.

[85] Law no. 2021-1104 of 22 August 2021 on combating climate disruption and strengthening resilience to its effects, JORF no. 0196 of 24 August 2021, art L.230-5-4.

[86] Conseil général de l'alimentation, de l'agriculture et des espaces ruraux. 2021. Évaluation de l'expérimentation du menu végétarien hebdomadaire en restauration collective scolaire, rapport n°20068.

[87] Conseil général de l'alimentation, de l'agriculture et des espaces ruraux. 2024. Mission sur l'évaluation de l'expérimentation de l'option végétarienne quotidienne pour les collectivités territoriales volontaires introduit par l'article 252 de la loi Climat et Résilience, rapport n°22106.

[88] Arrêté du 30 septembre 2011 relatif à la qualité nutritionnelle des repas servis dans le cadre de la restauration scolaire, JORF n°0229 du 2 octobre 2011.

[89] A scientific opinion by the National French Food Safety and Health Agency found that it was not appropriate to set a cap on the frequency of vegetarian menus in school cafeterias. Source: Agence nationale de sécurité sanitaire de l'alimentation, de l'environnement et du travail. 2021. Rapport d'appui scientifique et technique relatif aux fréquences alimentaires recommandées en restauration scolaire dans le cadre de l'expérimentation du menu végétarien, referral n°2019-SA-0205.

non-minced beef, veal, lamb or offal as the main course in at least four out of 20 meals" so as to guarantee proper iron intakes. The rule further requires that there should be four meals out of 20 containing fish meals and another maximum of four meals containing a preparation based on proteins of animal origin, thus undermining meat consumption reduction targets. Over the past few years, plans and recommendations[90] to revise the rule have stalled,[91] restricting the possibility of increasing the number of additional meat-free options per week. Accordingly, in a 2023 ruling, the Administrative Court deemed unlawful the municipality of Lyon's ordinance mandating meat-free menus in school cafeterias.[92]

Overall, laws and policies that aim to reduce meat consumption fall short of achieving nutritional, environmental, and climate goals. However, emerging policy initiatives addressing the food, health, and environmental nexus of meat production and consumption appear promising. An example of such policy in France is the National Strategy for Food, Nutrition and Climate,[93] which, as its name indicates, aims to bring several policies into alignment with one another, in an effort to increase coherence between agri-food, public health, and environmental policies. However, the strategy published in 2025 turned out to be disappointing due to the absence of binding regulatory measures aiming to support the transition to healthier and more sustainable diets. Furthermore, this strategy failed to include quantifiable meat reduction targets and vegetarian menus in food services.

3 Reform Proposals to Reduce Meat Consumption

The many negative externalities resulting from high levels of meat consumption have prompted technological food innovations aimed at replacing conventional meat, such as cultivated meat. However, the creation of a market for artificial meat replacements might not be as effective as initiating a transition toward meat reduction.

[90] Written question no. 3327 by Francesca Pasquini MP, 22 November 2022; Answer 31 January 2023.

[91] Abdoun, E. 2023. Cantines scolaires, les lobbys à la manœuvre. Que. Choisir, 29 May.

[92] Lyon Administrative Court, 23 March 2023, no. 2101388–2101390 (in French).

[93] Provided for in article 265 of the aforementioned "Climate and Resilience" Act.

3.1 *Substitution as a Solution: Cultivated Meat*

The general public first learned about cultivated meat when Dutch scientist Mark Post unveiled the first cultivated hamburger in 2013. The cost of this burger was then estimated at 250,000 euros. Cultivated meat is produced by growing animal cells in a controlled environment, where they are provided with nutrients (such as amino acids, glucose, vitamins, and inorganic salts, and supplemented with growth factors and other proteins) and conditions that encourage the cells to multiply and form muscle tissue.[94] Cultivated meat thus replicates the sensory and nutritional profile of conventional meat. Cultivated meat proponents present it as a sustainable solution to meet the global demand for animal protein, a demand set to increase by 60% by 2030.[95] The Intergovernmental Panel on Climate Change (IPCC) has also demonstrated that replacing conventional meat with cultivated meat could substantially reduce direct greenhouse gas emissions from the food production sector, provided that the energy used to produce cultivated meat is decarbonized.[96] Some animal advocates also support the development of cultivated meat as a way to reduce, or even eliminate, the need to exploit farmed animals.[97] However, because cultivated meat is not produced at an industrial scale, the climate and animal welfare benefits of this technology are still uncertain.[98]

From a legal perspective, the emergence of cultivated meat requires legislatures to strike a balance between regulating a new market and ensuring a high level of protection for consumers. Two specific legal challenges the EU legislature has dealt with are the adoption of a market

[94] The Good Food Institute. The Science of Cultivated Meat. Available at: https://gfi.org/science/the-science-of-cultivated-meat/. Accessed 22 December 2024.

[95] Quesne, A. 2022. Clinique juridique One Health—Une seule santé, Livre blanc sur l'agriculture cellulaire: les enjeux juridiques et éthiques de l'alimentation de demain. Available at: https://cjonehealth.hypotheses.org/1061.

[96] IPCC. 2022. Climate change 2022: Mitigation of Climate Change, Contribution of Working Group III to the Sixth Assessment Report of the Intergovernmental Panel on Climate Change, p. 1289 and 1322.

[97] For instance, see Vandebosch, M., and P. Lymbery. 2024. *Cultivated Meat to Secure our Future: Hope for Animals, Food Security, and the Environment*. Brooklyn: Lantern Publishing and Media.

[98] Chriki, S., Hallman, W., Hocquette, J.-F., Ellies-Oury, M.-P., and M. Takeuchi. 2025. Food culture and cell-culture: technical, ethical and social frontiers. *Npj Science of Food* 9(1), p. 49.

authorization procedure and the adoption of clear rules on the denomination of cultivated meat products.

As with terms to define plant-based products, conventional meat producers oppose the use of terms that they think should be reserved solely for conventional meat. Some conventional meat producers even seek to prohibit cultivated meat producers from using the word "meat" to define cultivated meat.[99] EU law does not provide clear marketing rules on the use of denominations and terms for cultivated food products, even though there exists a regulatory definition for "meat" in the 2004 Hygiene Regulation, as the "edible parts" of domestic ungulates, poultry, lagomorphs, and wild and farmed game.[100] However, in a 2024 ruling, the CJEU determined that "since 'meat' is defined, in point 1.1 of Annex 1, as 'the edible parts of animals,' a food not containing such parts cannot use the name 'meat,' even if that name is accompanied by details of the substitution of components or ingredients."[101] This ruling thus precludes producers from using terms such as "cell-based meat." At international level, the Food and Agriculture Organization (FAO) of the United Nations seems to have a preference for specifying that cultivated food products should be clearly designated as "cell-based," considering this formulation to be the most descriptive and neutral.[102] The FAO further recommends that consumers must be informed about the species from which the cells were extracted, given the possibility of allergic reactions identical to those caused by conventional meat products. Examples of recommended good practices by the FAO thus include the use of terms such as "cultivated chicken" or "cultured beef."

The EU also provides a market authorization procedure governed under the Novel Foods Regulation.[103] The scope of the Novel Foods Regulation encompasses cultivated meat since the Regulation defines

[99] For instance, in the US, the State legislature of Iowa adopted a law in 2024 that prohibits cultivated meat producers from using the word "meat" and any term associated with conventional meat products. Source: State of Iowa. 2024. An Act Prohibiting the Misbranding of Certain Food Products, and Providing Penalties.

[100] Regulation 853/2004 Laying Down Specific Hygiene Rules for Food of Animal Origin, Annex I, Point 1.1., 2004 O.J. (L 139) 55–205.

[101] C-438/23, *Protéines France, Union végétarienne européenne (EVU), Association végétarienne de France (AVF), Beyond Meat Inc. v Ministre de l'Économie, des Finances et de la Souveraineté industrielle et numérique*, 4 October 2024, para. 64.

[102] FAO. 2022. Food Safety Aspects of Cell-Based Food. Background Document One—Terminologies.

[103] Regulation 2015/2283 on Novel Foods, 2015 O.J. (L 327) 1–22.

novel foods as "any food that was not used for human consumption to a significant degree within the Union before 15 May 1997,"[104] and explicitly refers to cultivated meat defined as "food consisting of, isolated from or produced from cell culture or tissue culture derived from animals, plants, micro-organisms, fungi or algae."[105] The EU establishes a high standard of food safety by presuming that novel foods might pose safety risks on account of their innovative nature and because experience has not established that they can be consumed without risk by the population. As per the Novel Foods Regulation, novel foods are subject to a market authorization regime that aims to ensure that such foods do not present a safety risk to human health, taking into account the scientific data available.[106]

A 2023 report by the FAO and the World Health Organization (WHO)[107] provided an overview of the health and food safety risks posed by cultivated meat consumption, as well as an assessment of these risks in light of those associated with conventional meat production and consumption. In addition to the same risks associated with the consumption of conventional meat, scientists identify food safety hazards associated with the use of technologies, materials, and ingredients specifically used to produce cultivated food. Food safety risks include bacterial contamination, the presence of allergens, and the genetic stability of cells.[108] In France, a group of some 60 scientists also cautioned against the safety of cultivated meat, due to the scarcity of scientific data available to assess the health, environmental, and nutritional risks of this new food technology.[109] Additionally, the production of cultivated meat may involve the use of genetically modified organisms (GMOs), which can pose additional

[104] Ibid., art (2)(a).

[105] Ibid., art (2)(a)(vi).

[106] Ibid., art 7. See also Friant-Perrot, M. 2023. Innovation et sécurité alimentaire. In *Les 20 ans du règlement sur la législation et la sécurité alimentaire dans l'Union européenne—Bilan et perspectives*, Nathalie De Grove-Valdeyron and Marine Friant-Perrot, eds. 171–195. Bruxelles: Bruylant; Friant-Perrot, M. 2020. Analyse des risques liée aux nouveaux aliments. In *Innovation et analyse des risques dans le domaine de la santé et des produits de santé dans l'Union européenne. Regards croisés*, ed. Nathalie De Grove-Valdeyron, 151–176. Toulouse: Toulouse Capitole University Press.

[107] FAO and WHO. 2023. Food safety aspects of cell-based food.

[108] Ibid. The report mentions the risk of genetic (epi) drift in cell lines.

[109] Tribunes: L'essor de la "viande de synthèse" repose sur peu de données scientifiques. *Le Monde*, 25 February 2023.

specific risks and whose use is therefore strictly regulated in EU law.[110] According to a 2023 report by the French Senate, "…[food scientists] may have used [g]enetically modified organisms to produce recombinant proteins in the culture medium, which are essential for cell differentiation and proliferation. In addition, some companies have chosen to genetically modify the cell lines themselves."[111]

To date, the European Commission has not delivered any marketing authorization for cultivated meat, although companies have submitted applications.[112] As a result, the sale of cultivated meat products is not allowed in the EU. At national level, certain Member States have adopted legislation that aims to restrict market access for cultivated meat companies. For instance, France adopted a law in 2021 prohibiting cultivated meat in school cafeterias.[113] In 2023, Italy also voted in favor of a law that prohibits cultivated meat from being sold in the country.[114] However, these national measures will likely be preempted by EU law given that the Novel Foods Regulation provides a harmonized legal framework for the authorization of these products.[115] Other jurisdictions outside the EU, with laxer food safety standards, have authorized the marketing of certain cultivated meat products, starting with Singapore, which authorized the

[110] Regulation (EU) n°1829/2003/EC on genetically modified food and feed and Regulation 1830/2003 concerning the traceability and labeling of GMOs and the traceability of food and feed products produced from GMOs. Friant-Perrot, M. 2010. The European Union regulatory regime for genetically modified organisms and its integration into Community food law and policy. In *The Regulation of Genetically Modified Organisms-Comparative Approaches*, Luc Bodiguel and Michael Cardwell, eds. 79–100. Oxford: Oxford University Press.

[111] We translated French Senate. 2023. Rapport n°504 d'information fait au nom de la Commission des affaires économiques sur les aliments cellulaires, p. 92.

[112] Foote, N. 2022. In vitro meat companies prepare for first EU license applications. Euractiv, 8 April. For instance, Gourmey (France)—July 2024-EU application for cultivated foie gras and Mosa Meat (Netherlands)—January 2025-Application for cultivated beef fat.

[113] Article 230-5 paragraph 2 of the French Rural Code. Verderame, V.N. 2022. "Beef—It's Not What's for Dinner": The Implications of France's Newest Law for the Future of Labeling Cultured Meat in the European Union. *North Carolina Journal of International Law* 47(4).

[114] Disposizioni in materia di denominazione dei prodotti alimentari contenenti proteine vegetali (746) [Provisions concerning the designation of food products containing plant proteins (746)], Legge del 16 novembre 2023. (in Italian)

[115] Petetin, L. 2014. Frankenburgers, risks and approval European. *Journal of Risk Regulation* 5(2): 168–186. Hungary. 2024. Draft act prohibiting the production and placing on the market of laboratory-grown meat. 2024/0394/HU.

marketing of cultivated chicken in 2021. The US also authorized two companies to market cultivated meat in 2023.

Like the US and other jurisdictions, EU law also allows the marketing of artificial foods on its market, provided such food products do not present disproportionate food safety risks. However, beyond food safety considerations, there also exist issues of consumer acceptance.[116] For instance, the manipulation of living organisms raises ethical issues that might cause consumer mistrust, especially among vegetarian or vegan consumers. For instance, the use of fetal calf serum—obtained by slaughtering a pregnant cow and its fetus for use as a growth medium—is unlikely to be compatible with an ethical eating praxis. Moreover, labeling would not necessarily preserve consumer choice, given the inherent traceability shortcomings of this technology and the risk of fraud.[117] For these reasons, the French Senate, in a 2023 report, recommended that cultivated products not be authorized unless and until authorized under the marketing authorization procedure laid out in the EU Novel Food Regulation.[118]

Lastly, a more profound critique of cultivated meat stems from the opposition to the industrial agri-food production model. Food security advocates in particular have expressed concerns that cultivated meat, and alternatives to conventional meat in general, could aggravate the risk of monopolization in agricultural and food production. For instance, a 2022 report from the International Panel of Experts on Sustainable Food Systems (IPES-Food) stated that "[t]hese alternatives risk consolidating the domination of food systems by a few giants (...) promoting a Western diet rich in processed foods, leading to a loss of income for farmers in Southern countries, and reinforcing industrial supply chains that harm people and the planet."[119] These concerns are substantiated by the fact that the global meat corporations,[120] which already comprise a large

[116] Fishler, C. 1990. *L'homnivore*. Paris: Odile Jacob.

[117] Stephens, N., Di Silvio, L., Dunsford, I., Ellis, M., Glencross, A., and A. Sexton. 2018. Bringing cultured meat to market: technical, socio-political, and regulatory challenges in cellular agriculture. *Trends in Food Science & Technology* 78: 155–166.

[118] French Senate. 2023. Rapport n°504 d'information fait au nom de la Commission des affaires économiques sur les aliments cellulaires, p. 108.

[119] IPES Food. 2022. The politics of protein. Available at: https://www.ipes-food.org/_img/upload/files/PoliticsOfProtein.pdf.

[120] Such as JBS, Tyson Foods, Cargill, and WH Group.

concentration of the beef, poultry, and pork industries, also control 80% of the US market for cultured meat and plant-based alternatives.[121]

3.2 Transition Through Meat Reduction

Besides economic considerations, the transition to more sustainable diets must also ensure fundamental rights, particularly the right of everyone "to live in a healthy environment where biodiversity is guaranteed, in a world where animals are not treated cruelly."[122] In other words, the transition from meat-rich diets toward more plant-rich foods is a prerequisite to ensure the most basic fundamental human rights, such as the right to food, the right to a healthy environment, and the right to health. This underscores the need for the State to regulate economic freedoms adequately to protect the environment and public health.

A purely market-based approach will not be enough to achieve healthy, sustainable, and humane objectives in food systems. Specifically, it is unlikely that the protein transition will solely rely on the creation of a new market for meat alternatives. A profound reform of the rules applicable to animal-based products is also necessary so that these rules properly include environmental and health considerations for all, humans and nonhumans alike. For instance, the French legislature recognized the necessity to link "sobriety," thus implying the idea of reducing consumption levels, and "ecological transition." According to French law,[123] the ecological transition must be consistent with a new economic model that "goes beyond the linear economic model [...] by calling for sober and responsible consumption of natural resources and raw materials." In the food sector, this transition implies rethinking the interactions between the various elements of the food system, understood as "the way in which people organize themselves to produce, distribute and consume their food."[124]

[121] Gérard, M. 2022. Les "fausses viandes", des alternatives pas si vertueuses. *Le Monde*, 3 June.

[122] We translated Fabre Magnan, M. 2019. Et si les droits de l'homme étaient pris en compte dans le contrat? *Revue des droits et libertés fondamentaux* n°34.

[123] Article 110–1-1 of the Environmental Code.

[124] Malassis, L. 1979. *Economie agroalimentaire. Tome 1: Economie de la consommation et de la Production agroalimentaire*. Paris: Cujas; Rastoin, J.L. and G. Ghersi. 2010. *Le système alimentaire mondial: concepts et méthodes, analyses et dynamiques*. Paris: Quae; Lubello, P., Falque, A., and L. Temri, eds. 2016. *Systèmes alimentaires en transition*. Paris: Quae.

Such a reform must be rooted in economic law, which must be fully deployed to enable consumers to make more responsible choices; however, action is also required on environmental determinants through the regulation of the meat supply and its promotion.

One first solution to facilitate the transition toward more sustainable meat consumption habits would be to increase transparency on the market by requiring that producers and manufacturers disclose certain information to consumers.[125] This measure would be in line with Article 169, TFEU, which requires the Union "to promote the interests of consumers and to ensure a high level of consumer protection."[126] For instance, the creation of a food sustainability label would make the environmental impacts of animal agriculture more visible to consumers, and could therefore be a useful tool to help advance the transition to more sustainable diets. For this reason, the European Commission had announced the creation of "a sustainable labelling framework [covering], in synergy with other relevant initiatives, the nutritional, climate, environmental and social aspects of food products," as part of the 2020 Farm to Fork Strategy.[127] While this project stalled at EU level, the French legislature planned to adopt a food sustainability label through a 2021 law.[128] More specifically, on the issue of animal agriculture, the European Commission, in the Farm to Fork Strategy, had also announced the adoption of an animal welfare

[125] Also see Di Concetto, A. 2023. Farm Animal Welfare and Food Information for EU Consumers: Harmonizing the Regulatory Framework for More Policy Coherence. *European Journal of Risk Regulation* 1/2023; Di Concetto, A., and A.-S. Epstein. 2023. EU Consumer Information as a Tool to Regulate the Treatment of Farm Animals: Potential and Limits. *European Journal of Consumer Law*.

[126] TFEU, art 169.

[127] European Commission. 2020. Communication from the Commission to the European Parliament, the Council, the European Economic and Social Committee and the Committee of the Regions A Farm-to-Fork Strategy for a Fair, Healthy and Environmentally-Friendly Food System. COM (2020) 381 final, p 13. See also Di Lauro, A. 2023. La stratégie de la ferme à la table: vers plus de transparence sur les impacts écologiques des denrées alimentaires? In *Les 20 ans du règlement sur la législation et la sécurité alimentaire dans l'Union européenne—Bilan et perspectives*,Nathalie De Grove-Valdeyron and Marine Friant-Perrot, eds. p 254. Bruxelles: Bruylant.

[128] Law no. 2021–1104 of 22 August 2021 on combating climate disruption and strengthening resilience to its effects, JORF no. 0196 of 24 August 2021. The French administration has not yet adopted the specifications of this label. A technical consultation on the method for calculating the environmental cost of food products was launched by the French Ministry of Ecology on 29 April 2025. Available at: https://affichage-environnemental.ademe.fr/en/node/63.

label to inform consumers about the treatment of farmed animals and "better transmit value throughout the food chain."[129] Even though the European Commission's Platform on Animal Welfare had subsequently adopted the position that the animal welfare label should be voluntary,[130] a mandatory animal welfare label on all animal-source food products would have helped fill in the gap between consumers' preferences and citizens' values.[131] A mandatory animal welfare label would also help consumers reduce their meat intake by avoiding low-cost and cheap products. This, in turn, would incentivize farmers to adopt better animal welfare practices. However, like the food sustainable label, this project is at a standstill.

In the same vein, a second line of reforms would limit commercial speech, encouraging meat consumption by better regulating nutritional, health, and animal welfare claims on meat products.[132] The 2020 Farm to Fork Strategy contained a few promising measures, starting with the adoption of new rules on "nutrient profiles to restrict the promotion (via nutrition or health claims) of foods high in fat, sugars and salt,"[133] which therefore included meat. Such a measure was consistent with Regulation 1924/2006 on Nutrition and Health Claims Made on Foods (Health

[129] European Commission. 2020. Communication from the Commission to the European Parliament, the Council, the European Economic and Social Committee and the Committee of the Regions A Farm-to-Fork Strategy for a Fair, Healthy and Environmentally-Friendly Food System. COM (2020) 381 final, p. 8.

[130] Directorate General for Health and Food Safety. EU Platform on Animal Welfare. 2021. Conclusions of the animal welfare labelling subgroup of the EU Animal Welfare Platform. DOC.2021.07202. Available at: https://food.ec.europa.eu/system/files/2021-06/aw_platform_plat-conc_awl-subgroup-conclusion.pdf.

[131] Epstein, A.-S. 2023. Communication d'entreprise et bien-être animal. In *Les 20 ans du règlement sur la législation et la sécurité alimentaire dans l'Union européenne—Bilan et perspectives*, Nathalie De Grove-Valdeyron and Marine Friant-Perrot, eds. Bruxelles: Bruylant.

[132] In that sense, also see Di Concetto, A. 2023. Farm Animal Welfare and Food Information for EU Consumers: Harmonizing the Regulatory Framework for More Policy Coherence. *European Journal of Risk Regulation* 1/2023; Di Concetto, A., and A.-S. Epstein. 2023. EU Consumer Information as a Tool to Regulate the Treatment of Farm Animals: Potential and Limits. *European Journal of Consumer Law*.

[133] European Commission. 2020. Communication from the Commission to the European Parliament, the Council, the European Economic and Social Committee and the Committee of the Regions A Farm-to-Fork Strategy for a Fair, Healthy and Environmentally-Friendly Food System. COM (2020) 381 final, p. 12.

Claims Regulation),[134] which mandated the European Commission to "establish specific nutrient profiles and the conditions, including exemptions, which shall be respected for the use of nutrition and health claims on foods and/or categories of foods" by 2009. These nutrient profiles would thus form the basis for specific rules, including the reservation of certain nutrition and health claims regarding foodstuffs that comply with certain profiles. In practice, this would mean that manufacturers could not use terms such as "source of..." or "rich in protein" on products such as red or processed meat, on account of their high content of saturated fatty acids. Although the Commission eventually never adopted nutrient profiles rules, the legal basis for adopting these rules in tertiary law (administrative regulations) exists.

Despite the stalling of these nutrient profiles rules, a welcome development has been the adoption of specific rules that aim to better regulate environmental claims, through the adoption of Directive 2024/825, which amended the Directive 2005/29/EC on Unfair Commercial Practices,[135] to specifically prohibit misleading environmental claims.[136] The European Commission further proposed new rules which would specify the standards operators must meet when using environmental claims and labels, in a Proposal for a new Directive on Substantiation and Communication of Explicit Environmental Claims, published in 2023.[137] The scope of this new Directive, which is expected to be adopted in 2025, will likely include environmental claims on food products. These new rules come as a welcome development in light of the proliferation of misleading environmental claims on meat products. An example of such a misleading claim is the use of the label "climate-controlled" on factory-farmed pig products sold by Europe's largest pork producer, Danish Crown. With other plaintiffs, the Vegetarian Society of Denmark sued Danish Crown

[134] Regulation 1924/2006 of the European Parliament and of the Council of 20 December 2006 on Nutrition and Health Claims Made on Foods, art 4, 2006 O.J. (L 404) 14–15.

[135] Directive 2024/825 of the European Parliament and of the Council of 28 February 2024 amending Directives 2005/29/EC and 2011/83/EU as regards empowering consumers for the green transition through better protection against unfair practices and through better information, 2024 O.J. (L 2024) 825.

[136] For more on EU consumer law and animal protection goals, also see Chap. 6, Clémentine Baldon, "Corporate Social Responsibility and Animal Protection."

[137] European Commission. 2023. Proposal for a Directive of the European Parliament and of the Council on Substantiation and Communication of Explicit Environmental Claims (Green Claims Directive). COM/2023/166 final. Also see Chap. 6, Clémentine Baldon "Corporate Social Responsibility and Animal Protection."

for violating Denmark's consumer protection laws. In 2024, the Danish Western High Court ruled against Danish Crown for failing to substantiate its claim.[138]

While the European Commission has been proactive in protecting consumers from misleading environmental claims over the past few years, the EU institutions have not taken action to end humanewashing practices on food products. As a result, animal welfare claims are regulated broadly under Directive 2005/29/EC on Unfair Commercial Practices,[139] which prohibits misleading commercial practices defined as "a commercial practice [that] contains false information [...], deceives or is likely to deceive the average consumer, [...] and in either case causes or is likely to cause him to take a transactional decision that he would not have taken otherwise."[140] However, because animal welfare claims are not regulated under a special EU statute, and Member States do not always provide specific legislation either, vague and unspecific claims are allowed to proliferate in the EU.[141]

Another way of regulating commercial speech would be to provide stricter rules on the advertising of meat products, including on public health grounds, as the EU has done for tobacco and alcohol.[142] Such rules

[138] Niranjan, A. 2024. Danish firm's 'climate-controlled pork' claim misleading, court rules. *The Guardian*, 1 March. Available at: https://www.theguardian.com/environment/2024/mar/01/danish-firm-climate-controlled-pork-claim-misleading-court-rules.

[139] Directive 2005/29/EC of the European Parliament and of the Council of 11 May 2005 concerning unfair business-to-consumer commercial practices in the internal market, 2005 O.J. (L 149) 22–39.

[140] Directive 2005/29/EC of the European Parliament and of the Council of 11 May 2005 concerning unfair business-to-consumer commercial practices in the internal market, art 6(1), 2005 O.J. (L 149) 28.

[141] Di Concetto, A. 2023. Farm Animal Welfare and Food Information for EU Consumers: Harmonizing the Regulatory Framework for More Policy Coherence. *European Journal of Risk Regulation* 1/2023; Di Concetto, A., and A.-S. Epstein. 2023. EU Consumer Information as a Tool to Regulate the Treatment of Farm Animals: Potential and Limits. *European Journal of Consumer Law*.

[142] Delhomme, V.N. 2024. Minimum harmonization, experimentation and the internal market: balancing unity and diversity in the regulation of lifestyle-related health risks. In *Balancing Unity and Diversity in EU Legislation*, Ton van den Brink and Virginia Passalacqua, eds. 194–210. Cheltenham: Edward Elgar. Garde, A. and M. Friant-Perrot. 2014. The Regulation of Marketing Practices for Tobacco, Alcoholic Beverages and Foods High in Fat, Sugar and Salt- A highly Fragmented Landscape. In *Regulating Lifestyle Risks in Europe: Tobacco, Alcohol and Unhealthy Diets*, Alberto Alemanno and Amandine Garde, eds. Cambridge: Cambridge University Press.

could also take environmental protection as a legal basis given the deleterious effects of meat production on the environment. In that vein, the 2020 French Citizens' Climate Convention proposed policy reforms aimed at regulating the advertising of high-polluting food products.[143] The city of Haarlem, in the Netherlands, adopted a similar measure through a city ordinance banning meat advertising in public spaces in 2022.[144]

Lastly, a third line of reform could attempt to facilitate the purchase of more nutritional, sustainable, and humane food products. Many national and international organizations, such as the World Health Organization,[145] the EU[146] and the French authorities,[147] have produced reports listing policy interventions that aim to shape healthier food environments. For instance, a 2017 report by the French authorities recommended using regulatory measures to reduce marketing pressure and improve the nutritional quality of the food supply, including through nutrition-based taxation ("fiscalité nutritionnelle").[148]

[143] Citizens' Climate Convention. 2020. Les propositions de la Convention Citoyenne pour le Climat. Available at: https://propositions.conventioncitoyennepourleclimat.fr/pdf/tl/ccc-consommer-reguler-la-publicite-pour-reeduire-les-incitations-a-la-surconsommation-tl.pdf (in French).

[144] Boffey, D. 2022. Dutch city becomes world's first to ban meat adverts in public. *The Guardian*, 6 September. https://www.theguardian.com/world/2022/sep/06/haarlem-netherlands-bans-meat-adverts-public-spaces-climate-crisis.

[145] WHO. 2013. Global Action Plan for the Prevention and Control of Noncommunicable Diseases 2013–2020. WHA 66.10.

[146] With regard to the vulnerable public of children, see European Commission. 2014. Action Plan to Combat Childhood Obesity 2014–2020.

[147] Haut Conseil de la Santé publique. 2017. Propositions pour une Politique Nationale Nutrition Santé à la hauteur des enjeux de santé publique en France PNNS 2017–2021, collection des Avis et rapports du Haut Conseil de la Santé publique; Serry, A.-J., Escalon, H., and M. Friant-Perrot. 2020. Évolution des messages sanitaires devant accompagner les publicités de certains aliments et boissons et préconisations concernant l'encadrement du marketing des produits gras, sucrés, salés en direction des enfants; French Cour des Comptes. 2019. La prévention et la prise en charge de l'obésité; French Cour des Comptes. 2021. La politique de prévention en santé. Les enseignements tirés de l'analyse de trois grandes pathologies.

[148] Haut Conseil de la Santé publique. Propositions pour une Politique Nationale Nutrition Santé à la hauteur des enjeux de santé publique en France PNNS 2017–2021, collection des Avis et rapports du Haut Conseil de la Santé publique. For additional resources on the issue of animal source food and taxation, see Coller Animal Law Forum. Taxation. Available at: https://calf.law/factsheets/taxation. Accessed 22 December 2024.

On the one hand, taxing meat more could further discourage excessive consumption.[149] As identified by the European Commission in the EU Beating Cancer Plan,[150] Member States could modulate the national sales tax rates[151] to integrate nutritional considerations[152] and increase access to healthy, nutritious food.[153] On the other hand, Member States could increase the sales tax rate of meat products to discourage consumers from buying too much of it.[154] For example, the German government had considered raising sales tax on meat from 7% to 19%, but eventually decided not to go through with the proposal.[155] Similarly, the Danish government adopted a specific tax on saturated fats in 2011,[156] but eventually had to cancel it to facilitate inter-state trade with Germany and Sweden.

4 Conclusion

Limiting meat production and consumption levels is essential for several reasons: to protect animals, who suffer in industrial farm animal production systems; to safeguard consumer health against non-communicable diseases; and to preserve the environment, by reducing the negative externalities of such food production on air, water, soil, and biodiversity.

Given the scope of the challenge, effective meat reduction policies cannot rely solely on existing legal frameworks, or on a single regulatory

[149] Funke, F., Mattauch, L., van den Bijgaart, I., Godfray, C., Hepburn, C., Klenert, D., Springmann, M., and N. Treich. 2022. Toward Optimal Meat Pricing: Is It Time to Tax Meat Consumption? *Review of Environmental Economics and Policy* 16(2): 219–240.

[150] European Commission. 2021. Communication from the Commission to the European Parliament and the Council, European plan to beat cancer. COM(2021) 44 final, p. 13.

[151] As allowed in EU law, under Council Directive 2022/542 of 5 April 2022 amending Directives 2006/112/EC and (EU) 2020/285 as regards rates of value added tax, 2022 O.J. (L 107).

[152] Daudigny, Y., and C. Deroche. 2014. Fiscalité et santé publique: état des lieux des taxes comportementales. Rapport d'information n° 399, p. 103; Hammadi, R. 2016. Rapport d'information de la commission des finances, sur la taxation des produits agroalimentaires, n° 3868.

[153] In that sense, see also Conseil des Prélèvements Obligatoires. 2023. La fiscalité nutritionnelle, CPO note no. 5.

[154] Dufour, A.C. 2018. L'empreinte de la responsabilisation dans le financement de la santé. *Revue de droit sanitaire et social*, p. 1055.

[155] Perino, G. and H. Schwickert. 2023. Animal welfare is a stronger determinant of public support for meat taxation than climate change mitigation in Germany. *Nat Food* 4: 160–169.

[156] Tax of 16 crowns (€1.37) per kilo on foods containing more than 2.3% saturated fat (applicable to meat and animal fats in particular).

approach. While consumer information rules must be improved, increasing transparency on the food market will likely be insufficient to reach reduction targets. Information-based policies also tend to exacerbate social disparities, especially in an economic context of steep inflation, which may cause many consumers to choose low-cost products, despite the negative impact on their health and the environment.

A more horizontal, systemic approach will instead have much better chances for success in achieving the EU's touted goal of establishing "a fair, healthy, and environmentally-friendly food system."[157] Another challenge for policymakers will be to avoid creating cultural insecurity, given how politically charged meat is in meat-centered food cultures such as in Europe. In *Mythologies* (1957), Roland Barthes writes: "Like wine, steak is, in France, a staple, nationalized even more than socialized; it appears in all the settings of everyday food life." The path ahead is therefore complex. Update 10 july 2025

[157] European Commission. 2020. Communication from the Commission to the European Parliament, the Council, the European Economic and Social Committee and the Committee of the Regions A Farm to Table Strategy for a Fair, Healthy and Environmentally Friendly Food System. COM (2020) 381 final.

Index[1]

[1] Note: Page numbers followed by 'n' refer to notes.

A.-S. Epstein, A. Di Concetto (eds.), *Animals in EU Economic Law*, The Palgrave Macmillan Animal Ethics Series,
https://doi.org/10.1007/978-3-032-16171-0

GPSR Compliance

The European Union's (EU) General Product Safety Regulation (GPSR) is a set of rules that requires consumer products to be safe and our obligations to ensure this.

If you have any concerns about our products, you can contact us on ProductSafety@springernature.com

In case Publisher is established outside the EU, the EU authorized representative is:

Springer Nature Customer Service Center GmbH
Europaplatz 3
69115 Heidelberg, Germany

Batch number: 10755352

Printed by Printforce, the Netherlands